T0329823

Metaheuristics for Structural Design and Analysis

To Güler Toklu

Optimization Heuristics Set

coordinated by
Nicolas Monmarché and Patrick Siarry

Volume 3

Metaheuristics for Structural Design and Analysis

Yusuf Cengiz Toklu
Gebrail Bekdaş
Sinan Melih Nigdeli

WILEY

First published 2021 in Great Britain and the United States by ISTE Ltd and John Wiley & Sons, Inc.

ISTE Ltd
27-37 St George's Road
London SW19 4EU
UK

www.iste.co.uk

John Wiley & Sons, Inc.
111 River Street
Hoboken, NJ 07030
USA

www.wiley.com

Library of Congress Control Number: 2021932796

British Library Cataloguing-in-Publication Data
A CIP record for this book is available from the British Library
ISBN 978-1-78630-234-2

Contents

Preface

This book is about the use of metaheuristic algorithms on two different but closely related subjects, namely structural design and structural analysis. Indeed, these subjects are different, but every engineer knows that structural design involves structural analysis, most of the time applied many times in order to finish the design.

Applications of metaheuristic algorithms on design problems began as early as the first appearance of these algorithms, in the second half of the 20th century. In these works, analysis of parts were being performed by using the well-known finite element method (FEM) or some other techniques. In the literature, we can find an important accumulation of such applications, predominantly in scientific media. Despite the abundance of these works, engineers are still far from using them extensively in practice. The studies are still at the academic level except for certain problems, like some structural parts or components and some simple trusses.

The existence of FEM enabled engineers to easily and accurately analyze the behavior of common structures in a very large domain. But for some uncommon problems involving nonlinearities and discontinuities, the elegant FEM ceases to be sufficient and efficient. The use of energy principles combined with basic concepts of FEM – recently referred to as the Finite Element Method with Energy Minimization (FEMEM) – is shown to be able to solve these special problems with the help of metaheuristic algorithms. The studies in this field are also still at a primitive level, and far from being available in practice.

Thus, it can be said that applications of metaheuristic algorithms are not sufficiently developed for engineers, either in design or in analysis. But in both fields, the trend is for great advances to be made continuously, in every corner of the world, so that the subject is a highly dynamic one, and the goal to be attained is still way ahead.

Writing a book on a subject that is ever advancing in many aspects is generally rather difficult to do. Such an undertaking requires a certain kind of encouragement; for this book, that encouragement was provided by Patrick Siarry from Paris-Est Créteil University, France. The authors also thank those at Wiley-ISTE who were very helpful during each step of the preparation of the book. Special thanks go to Emeritus Professor Fuat Erbatur from Middle East Technical University, Ankara, Turkey, for introducing these algorithms to the authors back in 2000.

We hope that readers find this book to be a thorough evaluation of the state of the art for applications of metaheuristic algorithms in structural design and analysis.

Yusuf Cengiz TOKLU
Beykent University
Istanbul, Turkey

Gebrail BEKDAŞ
Istanbul University-Cerrahpaşa
Turkey

Sinan Melih NIGDELI
Istanbul University-Cerrahpaşa
Turkey

March 2021

Introduction

I.1. Generalities

Everything in our universe is related to some kind of optimization, minimization in losses and expenditures and maximization in gains. This is true for everything, from a stone or flowing water looking for the minimum potential energy position to living organisms trying to find the best solution when they come across a problem. As we all know from the laws of life, evolution is always towards the best fit, the word "best" pointing again to an optimization process.

In human life also, whether performed using their own intelligence or by artificial intelligence developed by them, optimization is everywhere: from engineering design to construction planning, from personal economics to world economics, from transportation to water supply, from space research to deep-sea analysis, from self-care activities to organization of hospitals, from art activities to educational systems.

Optimization is sometimes influenced by limited financial, physical and timely resources, sometimes influenced by certain intangible motivations like aesthetics and desire, and, most of the time, by a multitude of reasons. In real life, optimization concerning one single objective is very rare compared to multi-objective optimization. While performing optimization, we are usually bound by certain restrictions which are the constraints of the problem: the number of machines to be used in a production process, the magnitude of gravitational force that a pilot can withstand, or the strength limit of a given material. Thus, a general optimization problem can be formulated as

$$\min \mathbf{F(x)}, \mathbf{x} = ?, \tag{[I.1]}$$

$$\text{such that } \mathbf{g(x)} = \mathbf{0} \text{ and } \mathbf{h(x)} \leq \mathbf{0}, \tag{[I.2]}$$

where $\mathbf{F(x)}$ is a set of functions to be minimized called objective functions, $\mathbf{g(x)}$ is a set of equality constraints, $\mathbf{h(x)}$ is a set of inequality constraints and $\mathbf{x}$ is the set of unknowns over which all functions and constraints are defined. Optimization here is shown as minimization without losing generality, since a maximization function can be turned into a minimization function by just a simple multiplication with -1. The same is true between "smaller than" and "greater than" conditions.

There have been very elegant techniques using classical tools of mathematics for solving this optimization problem, in cases where functions in $\mathbf{F(x)}$ are well defined and differentiable. Unfortunately, all of these methods are valid only in their range of applicability, as in linear programming, nonlinear programming, integer programming, gradient methods, etc. (Nocedal and Wright 2006; Fletcher 2013). In real life, most of the objective functions are such that they cannot be written down as a mathematical function, let alone a differentiable one. On the other hand, the unknowns may be any type of quantity, floating or integer numbers, names, directional expressions, the order of some activities, etc. Therefore, we can surely conclude that mathematical optimization methods cannot handle these problems, which form the great majority of problems encountered in real life.

Metaheuristic methods, fortunately, are capable of handling all of these problems. Properly designed, they can help in making decisions on the best topology of a structural element, an economic activity with maximum income, the best hourly schedule in a school, the best route to follow between two points, etc. The term "metaheuristic" was first coined for the tabu search method, viewing it as "a metaheuristic superimposed on another heuristic" (Glover 1986). Humans started to use these techniques consciously in the 20th century, although nature was probably using them from the very beginning. For instance, evolutionary theory shows that living organisms of today started from single-celled beings, following the optimization rule of best fit, under an unimaginable number and variety of constraints. A genetic algorithm, a popular metaheuristic algorithm, is just an imitation of this process. Currently, there are hundreds of metaheuristic algorithms, as well as hybrid ones, that are applied to a wide variety of optimization problems in science, engineering, economics, arts, etc.

Metaheuristic algorithms are also considered to be one of the most useful tools of artificial intelligence, taking into account that self-learning and rule-of-thumb decisions are their two basic properties.

This book is intended to give a review of metaheuristic algorithms and their applications in a very specific field: structural design and analysis. It is to be noted that this is the first book to deal with the application of metaheuristic algorithms to structural analysis.

I.2. Structure of the book

This book is organized in the following manner.

Chapter 1 gives a short history of structural analysis and design, from the times when these activities were performed using intuition and experience, without making any calculations, to times when tools used in artificial intelligence became frequent applications. This chapter emphasizes that the finite element method (FEM) plays a special role, whilst also noting that every step in this long voyage had a certain importance.

Chapter 2 gives an overview of metaheuristic algorithms (MAs). These algorithms started to be consciously used in the second half of the 20th century, enabling optimization problems to be solved that were untouchable before that time. In the beginning, there were only a handful of MAs, now the number certainly runs into the hundreds. In this chapter, some general properties of all of these algorithms are discussed, and about ten are investigated in detail.

Metaheuristic algorithms are successfully applied to structural problems. A general overview of these applications is given in Chapter 3, with emphasis on various aspects of the aims of optimization, i.e. the objectives. Examples are given in terms of weight, cost, effectiveness optimization, minimization of CO_2 emissions and dealing with limitations of stresses, deformations, stability, fatigue and national and international specifications.

The following four chapters are dedicated to design optimization. Generalities about applications of metaheuristic algorithms on structural design are discussed in Chapter 4. Chapter 5 deals with trusses and truss-like structures, Chapter 6 focuses on optimization of structural elements, and optimization of structural control members is the subject of Chapter 7. In all three of these chapters, after providing basic information about the subject, relevant numerical examples are given. Thus, in this part, the reader can find solutions for the optimization of an I-beam, a tubular column, a cantilever beam, trusses with elements – in which the number changes between 5 and 200 – reinforced concrete members, frames, walls and tuned mass dampers.

As stated in Chapter 3, optimization procedures can be useful not only in structural design but also in structural analysis. This subject is addressed in Chapter 8. The idea of these applications is the direct use of the minimum potential energy principle of mechanics in determining the equilibrium position of a structure. It is explained in the chapter that a method, named Total Potential Optimization using Metaheuristic Algorithms (TPO/MAs), was launched for this purpose, which can also be looked upon as Finite Element Method with Energy Minimization (FEMEM). In the chapter, the fundamentals of the method are given, together with applications on trusses, truss-like structures and plates.

Applications of metaheuristic algorithms on the design and analysis of structures are a relatively new subject with advances made every year and in every corner of the globe. In the concluding chapter, future expectations on this subject are discussed. It is stated that the tools used nowadays are basic tools of Artificial Intelligence (AI) and that with the amalgamation of design and analysis – along with other aspects of construction like management, planning, financing, controlling and site work – a huge problem lies ahead, requiring much more elaborate tools in order to be solved. When one considers that these operations will not only be carried out in familiar environments, but also perhaps in remote areas with harsh conditions, the difficulty of the task awaiting humanity can somehow be envisaged.

1

Evolution of Structural Analysis and Design

Design and analysis are two extremely important aspects of structural engineering. Although they are known to be applied side by side, or one inside the other, historically, the case was different. Design began in the very early ages, together with the first human settlements, while analysis only saw the light of day during the Age of Enlightenment. In early times, the design of structures was accomplished by intuition, observation, experimentation and experience, as can be seen in the applications of ziggurats, pyramids, St. Sophia in Istanbul, Notre Dame in Paris and other historical structures of varying sizes.

Structural analysis involves concepts such as forces, bending moments, torques, stresses, strains, deformations, deflections, rotations and warping, which are difficult to comprehend for those who are not educated in this specific subject. Also, it is evident that comprehension of these concepts is not sufficient to produce any type of analysis; it has to be enriched by knowledge in order to conceive a mathematical model, after making some meaningful assumptions, as well as computational experience to solve multiple equations or to find the minimum of a function. Hence, analysis, which we know today to be part of design, was developed much later than the design that was sufficient to build many important structures like the ones cited above. Of course, now humans can design structures that were previously in conceivable as a result of our improved powers of analysis, amongst other factors.

1.1. History of design

Designing a structure requires that certain questions be answered in relation to safety, functionality, economy, sustainability, aesthetics and comfort. The first question, that of safety, is directly related to structural analysis. It involves characterizing the materials to be used in the structure, making a mathematical model of the structure after making some meaningful assumptions, giving meaningful dimensions to the elements of the structure, determining design forces with a distinction between dead and live forces and calculating internal effects in the structures under different loadings, so as to support reactions, stresses, strains, deflections and rotations. Then comes the stage for determining the safety of the structure, i.e. deciding whether the structure, whose behavior has been calculated, is sufficiently strong to resist the acting forces with a meaningful factor of safety, whether the stresses are less than the materials can safely carry and whether the deflections and rotations are not severe enough to create a sense of discomfort. If any of the answers to these questions are negative, then either the dimensions of the structure will be changed or, in some cases, the entire model will be changed as a more radical measure. A change in the dimensions may be in two directions: increase the dimensions if the element is overstressed, or decrease them if the element is highly under-stressed to avoid having a structure that is far from economical.

The flowchart of this process is shown in Figure 1.1. It should be noted that the algorithm represented in this chart is not fully applicable, even in our age, though engineers are close to using it only for some simple structures. We can estimate that full use of the algorithm shown will be possible with advances yet to be made in the field known as artificial intelligence.

In the early ages, there was of course no knowledge about concepts such as stress, strain, force, deformations, stability, buckling, bending and torsion. Formulating a mathematical model of the structure was unthinkable, and there was no knowledge of calculus to do any kind of computation. Despite all of this, there was the intuition, experience, intelligence and observation power of some extraordinary people.

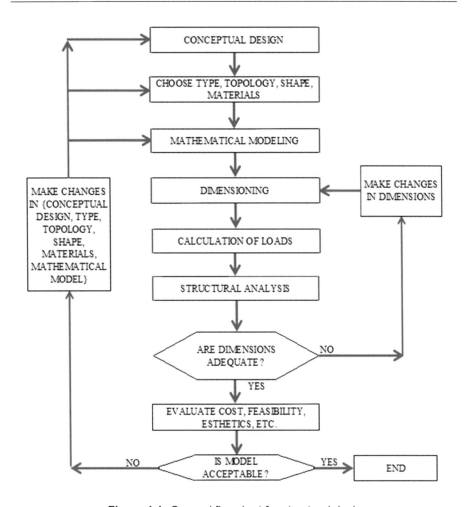

Figure 1.1. *General flowchart for structural design*

In the beginning, the above algorithm comprised only the first step shown; the engineers/architects started building structures based only on conceptual design: a temple, a ziggurat, a pyramid, an obelisk, etc. and made improvements to their designs, according to their feelings, as construction went on. The most important characteristic of the flowchart shown above is the existence of iterations at various steps. It is easy to assume that iterations during these early times were being performed in the field, i.e. during the construction process. Perhaps tens of years or more were necessary for field iterations for the cutting of stones; carrying them to their places, carving figures wherever necessary, placing small stones and erecting T-shaped columns at Göbeklitepe – the first known edifices of humankind, recently uncovered at Urfa, Turkey – built some 12,000 years ago (see Figure 1.2).

Figure 1.2. *Representative drawing showing construction operations at Göbeklitepe, Urfa, Turkey (https://www.cnnturk.com/kultur-sanat/dunyanin-en-eski-tapinagi-gobekli-tepe?page=1). For a color version of this figure, see www.iste.co.uk/toklu/ metaheuristics.zip*

We also see field iterations in structures that have collapsed and been rebuilt. The best example that can be given from those years is the Hagia Sophia (see Figure 1.3) in Istanbul, Turkey, which was constructed between February 532 and December 537, following the design created by Anthemius of Tralles and Isidorus of Miletus (Mainstone 1988; Ozkul and Kuribayashi 2007; Çakmak *et al.* 2009). This design was carried out using the experience and intuition of the two architects, perhaps with some computations with respect to the sizes of the columns. It was not possible to perform the computational iterations depicted in Figure 1.1, but they forcedly took place to end up with a safe structure. In May 558, i.e. almost 20 years after construction had finished, "parts of the central dome and its supporting structural system collapsed" (Çakmak *et al.* 2009). Those failures were probably due to earthquakes that took place in August 553 and December

557. Subsequently, a new dome – approximately 6.24 m higher than the original – was designed by Isidorus the Younger, nephew and successor to Isidorus of Miletus. This can be considered as an iterational process on the conceptual design of the structure, performed in the field as a real construction.

Figure 1.3. *Current photo of Hagia Sophia in Istanbul, Turkey. For a color version of this figure, see www.iste.co.uk/toklu/metaheuristics.zip*

The real iterations started with advances in structural analysis; its history written by Timoshenko (1983), Benvenuto (1990), Mainstone (1997), Felippa (2001), Addis (2003), Kurrer (2018) and many others.

Strength of materials is a primordial field in structural analysis. Following an introduction that charts its early beginnings, Timoshenko (1983) – in his book originally published in 1953 – explores the history of this subject across 14 chapters, beginning with the 17th century and ending with the period 1900–1950.

Mainstone (1997) states that structural analysis, as we now know it, began in the 18th century to assess the safety of buildings to be constructed. By the mid-19th century, this area is extended for analyzing earlier structures like Gothic cathedrals. The first idealization was Hooke's law of direct proportionality; this idealization went together with increasing knowledge about statically determinate structures, ignoring those that were statically indeterminate or hyperstatic. The second half of the 19th century was marked by graphical methods.

Matrices hold a particularly important place in the history of analyses of structures. Felippa (2001) addresses this aspect under three titles: creation, unification and FEMinization, the last term meaning the launching of "[...] the direct stiffness method [...] as an efficient and general computer implementation, as yet unnamed, finite element method (FEM)". Addis (2003) states that the main causes of failure in historical structures were wind loads, foundation problems and fires. We must add earthquakes to these effects for some parts of the world. Addis mentions the importance of full-scale tests on small physical models in historical times.

Two books, authored by Benvetuno (1990) and Kurrer (2008), are especially remarkable in following the history of structures. In both, the subject is traced from the very first works up until the time each book was written, dividing the advances into eras with historical drawings and anecdotes. Benvetuno's book is in two parts: "Statics and resistance of solids" and "Vaulted structures and elastic systems". Kurres's book, titled *The History of the Theory of Structures – From Arch Analysis to Computational Mechanics* was published originally in German, in 2002.

Although the design, analysis, and construction of structures have always shown continuous advances in history, there was a period when this was the opposite. Indeed, between the time of the Romans and the Renaissance, let aside making of constructions more important than in previous times, the knowledge accumulated until then was lost (Timoshenko 1983).

1.2. From empirical rules and intuition to FEM

In the beginning, the design of all types of edifices was being carried out without any form of analysis since structural calculations were not available by any means. Due to the field iterations mentioned in the above sections, some empirical rules were being brought forward, which were, mostly geometric (Cowan 1977). An example of these rules is the one about vaults (see Figure 1.4) noted by Leonardo de Vinci (1452–1519). This stated that the chords connecting the top of the structure to the extreme points at the bottom should not cross the inner arch, in order to prevent failure (Benvenuto 1990, p. 9).

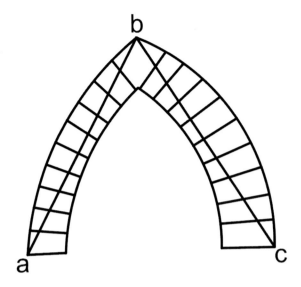

Figure 1.4. *Empirical rule on vaults: lines ab and bc will remain within the wall*

Empirical rules were in use until the 18th century. As time went on, structural computations began to come onto the scene. One of the starting points of this trend, advanced by Robert Hooke (1635–1703), was the rule "*ut tensio, sic vis*", "extension is directly proportional to force", which was the starting point of the theory of elasticity. We can assume that iterations of structures have begun, first in terms of dimensions, second in terms of shapes, with the introduction and advent of structural computations. But since the structural theory was only at a primitive level, and computational tools were not available, the iterations were very naïve at the beginning. A method known as the cross method, advanced in 1930 by Hardy Cross (1885–1959), which is technically known as the "moment distribution method", became a very important tool for engineers in structural computations (Cross 1930; Eaton 2001). The introduction of computers in 1960 enabled designers to do their computations in a much more efficient way. Finally, the FEMinization, defined above, being advanced towards the end of that decade (Felippa 2001), enabled engineers to analyze all types of structures in a very efficient way, eliminating the fear of a higher number of unknowns and of complex structures.

With these appropriate tools and theory, engineers started to make efficient iterations in designing structures, since it was now possible to repeat analyses easily by changing the dimensions of the members and even their placements.

1.3. From FEM to AI

The cross method and those similar, like the slope deflection method, were to be applied to beams and frames only; thus, they had a limited area of application. Other types of structures were being analyzed all by special methods, based on advanced (at the time) theories on elasticity, strength of materials, plates and shells. FEM, in a very short time after its introduction, became capable of handling all types of structures, including trusses, beams, frames, plates, shells, grids, volumes and any combination of them. Except for some special structures, the CPU (central processing unit) time spent for a normal structure was negligible. CPU time was becoming important only when the number of unknowns of the problem at hand was immense and also when there were nonlinearities or other particular characteristics. Therefore, FEM suddenly became the unique tool in engineers' hands for analyzing buildings, dams, bridges, etc. This, of course, enabled engineers to make iterations on sizes and also on shapes and topology. In fact, making iterations on sizes, i.e. on dimensions, can be done much more easily in comparison to the other two. Making iterations on shapes and especially on topology is much more delicate since changing them may result in also changing many other parameters, the effects of which cannot be foreseen easily.

Reaching this level enabled engineers to make optimizations of structures far better than the time before FEMinization. Choi *et al.* (2016) cite the work of Schmit (1960) as the initiator of structural optimization with systematical means. Schmit enumerates three steps in this process:

1) establish a trial design;

2) carry out an analysis based on this trial design;

3) and, based on the analysis, modify the trial design as required.

In this computerized time before the introduction of FEM, Schmit used a systematic way of optimizing a structure under different loadings, without any user intervention, on an example of a truss with three bars. In fact, establishing a structural optimization process for a general type of structure is not a stage that has been reached, even in our times. There are several commercial software packages actually on the market. Three of them are compared to one another on several structural optimization problems (Choi *et al.* 2016). In these packages, analysis is performed by FEM and optimization is carried out by nonlinear optimization methods.

More recent studies on the design of structures make use of metaheuristic algorithms (MAs) in the optimization part of the process (see, for example, Toklu (2009); Ahrari and Atai (2013); Saka *et al.* (2016); Techasen *et al.* (2019)). Indeed, MAs are much more versatile than any other mathematical optimization technique, they accept any type of state variables, they can deal with constraints and multiple objectives very easily, and nonlinearity imposes no difficulty on them (Gandibleux *et al.* 2004; Sirenko 2009; Boussaïd *et al.* 2013; Collette and Siarry 2013; Toklu and Bekdaş 2014; Sorensen *et al.* 2018; Almufti 2019).

In most applications, optimization is applied in design for determining the most appropriate structure, as far as topology, shape and size is concerned, analysis being performed by FEM as mentioned above. In some recent applications, analysis is also carried out by optimization processes, through a technique called Total Potential Optimization using Metaheuristic Algorithms (TPO/MAs) (see, for instance, Toklu (2004a); Toklu and Uzun (2016); Toklu *et al.* (2017); Bekdaş *et al.* (2019a); Nigdeli *et al.* (2019); Kayabekir *et al.* (2020a); Toklu *et al.* (2020)). TPO/MA is nothing but the application of the FEM method with an optimization process using soft computing methods, instead of solving matrix equations; thus, it deserves to be called the Finite Element Method with Energy Minimization (FEMEM) in a more general way.

The use of metaheuristic algorithms in the analysis of structures, together with topology, shape and size optimization, takes the structural design problem to a very different level. The importance of stochastic algorithms in structural design has already been noted by Kress and Keller (2007). Artificial intelligence (AI), as first imagined by Alain Turing (1912–1954), necessitates computing machines "(1) learning from experience and (2) solving problems by means of searching through the space of possible solutions, guided by rule-of-thumb principles" (Copeland 2000). Metaheuristic algorithms are exactly the ones guided by rule-of-thumb principles. Thus, we can easily say that structural design, with all its components, belongs now to the area known as AI (Lu *et al.* 2012; Yang 2013; Yang *et al.* 2014a; Hao and Solnon 2020).

2

Metaheuristic Algorithms

In engineering there are some complex problems that are impossible or very difficult to solve using classical methods. In those cases, several approximations and simplifications become necessary in order to arrive at a solution. But obviously, these arrangements must ensure that the results obtained are acceptable and close to reality. Numerical methods may be considered as a remedy in these cases. Metaheuristic-based algorithms may be used to increase the reliability of the solutions to the complex problems. In fact, these algorithms, which are stochastic in nature, can handle all types of variables, functions and constraints, and can determine the best solutions after making a number of iterations, finding their way with rule of thumb decisions.

Generally, the most challenging engineering problems are the optimization of designs. In civil engineering (especially in structural engineering), the number of design variables is more than other engineering disciplines, and the design constraints are dependent on several design variables. For this reason, the problems are highly nonlinear.

The design variables are considered differently in continuous and discrete optimization. In continuous optimization, the design variables are real numbers and exact optimum results are found. In several engineering applications, discrete design variables are used. This is especially the case in structural design, in which the dimension or size values may not be provided in real number values. In this case, the design variables are integer and the values are fixed according to the market sizes of materials and the practical production of structural members in size. Also, the design variable may be a string. Another option is to use mixed design variables, as some of the design variables must be discrete while continuous variables can be accepted for several.

The application area of metaheuristics must not only be treated as the optimization. For nonlinear analyses of structural systems, metaheuristics can also be used by minimizing the total potential energy of structural systems (Toklu 2004b).

Generally, metaheuristic algorithms are nature-inspired. In nature, several processes are instinctive. These processes are mathematically written as algorithms for minimizing or maximizing objectives, by considering physical, desirable and visual-based design constraints.

2.1. A brief history of the development of metaheuristic algorithms

Day by day, the number of metaheuristic algorithms increases. The newly developed methods sometimes outperform the existing algorithms for specific problems. The use of variants from existing methods, such as classical ones, is an active area of research, since a metaheuristic method may be best for one type of problem, while it is not the best choice for another type of problem. In that case, a problem must be tested for several algorithms and compared with the others, by considering statistical performance and computational time. This situation is called the "no-free-lunch" theorem.

The periods of development of metaheuristics are given in five groups, such as the pre-theoretical period (until c.1940), the early period (c.1940–c.1980), the method-centric period (c.1980–c.2000), the framework-centric period (c.2000 to now) and the scientific period (the future), by Sörensen *et al.* (2018).

In the pre-theoretical period, the human brain used heuristics; however, it is not formally studied. Humans need optimization. The basic optimization problems are essential problems, such as the finding the shortest path problem, patching problem and knapsack problem. From early childhood, the human mind is formidably equipped to solve a great range of problems, and is capable of using metaheuristic strategies. For example, when a human needs to solve a new problem, similar solutions to past-solved problems are found, and the rules are derived by solving this problem. As a final remark from Sörensen *et al.* (2018), heuristics and even metaheuristics are natural to humans.

After World War II, a book called "How to Solve it" was published by Polya (2004). In this book, it is emphasized that problems can be solved by a limited set of generally applicable strategies to make the problem easier. The focus of this book was not optimization; however, the proposed strategies are applicable to developing optimization algorithms. The early period started in 1940, and several principles that are useful in the design of heuristic algorithms are mentioned. The principles are called analogy, induction and auxiliary problems. The analogy principle proposes looking at another solved problem that closely resembles the current problem. The induction principle is to solve a problem by using a generalization of some examples. A sub-problem can also be used in solving an auxiliary problem. In metaheuristics, the problem that is trying to be solved is studied by another problem (analogy), and the learned techniques are used (induction). During solving, the problem is decomposed into smaller sub-problems (auxiliary problem). Also, greedy selection rules that select the best value at each iteration, (Kruskal's or Prim's algorithm for the minimization of spanning tree problem, and Dijkstra's algorithm for the shortest path problem (Simon and Nevell 1958)) are proposed in this period. Simon and Nevell (1958) also mentioned "ill-structured" problems that cannot be formulated explicitly or solved by existing methods easily.

In the method-centric period, the earliest frameworks of metaheuristics were developed. These early metaheuristics, such as the tabu search, were developed by Glover (1986) after some applicable strategies and inspiration for the development of optimization algorithms. During this period, the algorithms were mentioned as artificial intelligence methods since the methods involved mimicking human problems with solving behavior from learning lessons. These methods used life's main problem of evolution by natural selection, in which Darwin linked the evolving process of species to their adaptation to their environment. Evolutionary algorithms were first developed during the 1960s; however, the true dates of conception for evolutionary algorithms are considered to be 1975 (Holland 1975) and 1986 (Goldberg and Samtani 1986). The oldest evolutionary algorithm is the Genetic Algorithm (GA). The classical form of GA uses coding of values with bits. Currently, numerical values are generally used.

Another metaphor, which is one of the first, is the annealing process, in which controlled heating and cooling processes are used in materials to prevent defects. Simulated annealing uses random solutions and accepts solutions if they improve. Different from the new generation methods, simulated annealing uses a single vector instead of a series of vectors for a population (Kirkpatrick *et al.* 1983). The fundamental methods were generally proposed in the method-centric period.

Then, the framework gains more popularity instead of the methods in the millennium. This period is called the framework-centric period. In this period, the number of metaheuristic algorithms showed a great increase, including hybrid metaheuristic methods. During this period, researchers worked on more than one framework for hybrid algorithms.

After the 1980s, a sub-period, called the metaphor-centric period, occured (Sörensen *et al.* 2018). Several metaheuristic algorithms named with the inspired metaphor have been developed and are still in development. Developed by several researchers, these newly proposed methods are criticized (Weyland 2010; Sörensen 2015) because of a lack of science. The future period will be science-oriented, and the metaphor will not be the main novelty.

2.2. Generalities about metaheuristic algorithms

All metaheuristic algorithms have specific features. For these features, the use of a new metaheuristic is an active research area to find the best suitable method, by searching for the best solution, computation effort and sensibility.

Among these specific features, two major components exist in all metaheuristic algorithms. These are randomization and the selection of the best solution (Yang 2010a). By the selection of the best solution, a good convergence to a local best solution leads designers to a local optimum value, which is not the true solution. In structural analyses using metaheuristic methods, especially, the local results are generally not useful as they correspond to equilibrium configurations that are of secondary importance.

The randomization content prevents trapping to a local optimum. A good balance between randomization and the selection of the best solution must be provided.

Generally, the specific features of metaheuristic algorithms are used after the generation of initial solutions. The generation of a design variable (x_i) is done using a solution range of the variable with lower (x_{min}) and upper (x_{max}) bounds, as formulated in equation [2.1]. ε is a random linear distribution that is randomly generated between 0 and 1. Also, the same design variables are generated several times. Generally, the number of generations of a design variable is referred to as a "population". The "population" can gain more specific names like the number of pollens, harmony memory size, the number of bats in flower pollination algorithm, harmony search algorithm and bat algorithm, respectively. These algorithms are population-based algorithms. Also, a single agent or solution is considered in trajectory-based algorithms, such as simulated annealing.

$$x_i = x_{min} + \varepsilon(x_{max} - x_{min}) \qquad\qquad [2.1]$$

In a population-based algorithm, the steps of the design methodology are as follows. A general flowchart is presented in Figure 2.1.

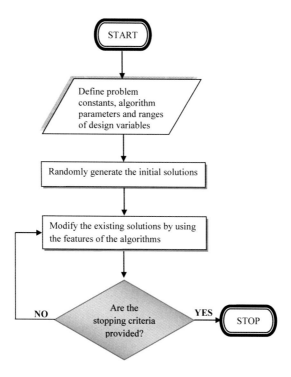

Figure 2.1. *General flowchart of metaheuristic algorithms*

Step I: define the problem constants, algorithm parameters and ranges of design variables.

Step II: randomly generate the initial solutions.

Step III: modify the existing solutions by using the features of the algorithms.

Step IV: continue iterative Step III until the stopping criterion (or criteria) is met.

In the modifications of existing solutions, most of the metaheuristic algorithms have two phases, which are exploration and exploitation. In exploration, the existing results are modified with a global range. This is generally called global optimization. In this phase, the possible minimum or maximum values distributed in all of the solution ranges are scanned. In global optimization, the convergence speed and precision in the optimum results may not be provided. In the second phase, an exploitation phase named local optimization is used to scan the areas close to the best or existing solutions. However, by using the exploitation phase too much, the optimization process may trap a local result, while a better result is found in another part of the solution range.

The ratio of usage of exploration and exploitation phases is decided according to a specific parameter of algorithms. The modification of the existing values in these phases is also done by using specific algorithms. For that reason, an algorithm with a less specific parameter is the most user-friendly, and these algorithms are generally chosen to skip the parameter tuning process.

For a specific engineering problem, while the exploration phase of an algorithm is effective on the problem, another algorithm may be effective in the exploitation phase. In that case, these algorithms are hybridized to eliminate the disadvantages. The newly generated hybrid algorithm may serve all of the good advantages for the problem.

In this chapter, several metaheuristic algorithms are briefly summarized; these algorithms are categorized as evolutionary algorithms, swarm intelligence and other metaheuristic algorithms.

2.3. Evolutionary algorithms

In the development of metaheuristic algorithms, the use of biological evolution mechanisms, such as reproduction, mutation, recombination and selection is the most well-known technique, and these algorithms are grouped as evolutionary algorithms. In this class, the most popular and oldest algorithms are genetic algorithms. For that reason, the other newly generated evolutionary algorithms are referred to as variants of genetic algorithms. In this section, genetic algorithms and differential evolution are briefly explained. Zhou *et al.* (2011) presented a survey on developed applications of multi-objective evolutionary algorithms.

2.3.1. *Genetic algorithms*

Genetic algorithms (GA) are one of the oldest metaheuristic algorithms that are still active in engineering problems. John Holland used crossover, mutation and selection in line with the evolution theory of Charles Darwin (Holland 1975). Goldberg and Samton (1986) first used GA for structural optimization by solving a 10-bar truss structure optimization problem. With the development of the new generation algorithm, the classical form of GA was outperformed; however, new variants of GA and the implementation of the features of GA on other algorithms may still be effective on several engineering problems.

In GA, encoding and decoding are needed, while the objectives are directly solved in the new generation methods. For that reason, GA is a modification tool for the other algorithms and uses the crossover, mutation and selection features. When the newly developed algorithms are scrutinized, these features are also used in the formulation of these algorithms. For example, in the local optimization phase of algorithms, generally, two different existing solutions are chosen, and the new modification is carried out. The selection of the two solutions is the selection, and the use of these solutions in one equation is the crossover. Then, the generated result is the mutation. For that reason, GA can be seen as the basis of metaheuristic algorithms. Srinivas and Patnaik (1994) provided a survey on the development of GA due time.

The steps of GA are listed as follows:

– encoding of objectives;

    – definition of fitness function or selection criterion;

    – initializing a population;

    – evolution of fitness function;

    – generation of a new population with the use of crossover, mutation and reproduction;

    – evolving the solutions until the stopping criteria are provided;

    – decoding the results.

2.3.2. *The differential evolution algorithm*

Storn and Price (1997) developed a metaheuristic algorithm called differential evolution, by using the mutation, crossover and selection features. The purpose of the development of the algorithm is to satisfy the following four demands of users.

First, the algorithm must be suitable to handle non-differentiable, nonlinear and multimodal objective functions. Second, the algorithm must be able to cope with the analyses of the intensive objective functions involving the design stages. Third, the algorithm must be user-friendly. For that reason, the control parameters must be less in number, and these parameters must be robust. The final demand is a good convergence property to save computational time.

After the generation of the initial solutions with N population, a mutated vector is generated as follows.

$$v_i^{t+1} = x_{r1}^t + F(x_{r2}^t - x_{r3}^t) \qquad\qquad [2.2]$$

The mutated ith solution, v_i^{t+1}, is generated by using three random existing solutions: x_{r1}^t, x_{r2}^t and x_{r3}^t. F is a control parameter for the amplitude of $x_{r2}^t - x_{r3}^t$, and it is a real constant number between 0 and 2.

Then, crossover is performed by using the existing $(x_1^t, x_2^t, \dots, x_N^t)$ and mutated $(v_1^{t+1}, v_2^{t+1}, \dots, v_N^{t+1})$ solutions according to equation [2.3], and the trial solutions $(x_1^{t+1}, x_2^{t+1}, \dots, x_N^{t+1})$ are generated:

$$x_i^{t+1} = \begin{cases} v_i^{t+1} & \text{if rand}(1) \leq CR \text{ or } i = r \\ x_i^t & \text{if rand}(1) > CR \text{ or } i \neq r \end{cases} \qquad [2.3]$$

rand (1) is a randomly generated real number between 0 and 1. CR is the crossover constant, which is the accepted percentage of the mutated solutions, and it is assigned between 0 and 1. r is an index, which is a randomly chosen integer number between 1 and N. By choosing a random index, it is ensured that at least one of the trial solutions is mutated.

In the selection section of the algorithm, the trial and existing results are compared by means of the objective functions. The best one is stored and the best solution is used as the existing solution of the next iteration. The variants and applications of differential evolution are detailed in a survey by Das and Suganthan (2011).

2.4. Swarm intelligence

The behavior of swarms is imitated in algorithms using swarm intelligence. In nature, every living thing has a specific processes to find food or prey, reproduce and sustain their species. These processes have a final objective, as we need a precise result in engineering problems. For that reason, the algorithms based on swarm intelligence are suitable and successful in solving structural engineering problems.

There are many types of swarm intelligence-based metaheuristic algorithms. Particle swarm optimization (PSO), developed by Kennedy and Eberhart (1995), is the best-known algorithm. The movement of members in a flock of birds or a school of fish is formulated in PSO.

New generation metaheuristic algorithms are still in development, while the classical ones are modified or combined with features of other metaheuristic algorithms. The inspiration of swarm intelligence-based algorithms is: finding the path between the colony and food in the ant colony optimization (ACO) (Dorigo *et al.* 1996), the attractiveness and the flashing characteristic of fireflies in the firefly algorithm (FA) (Yang 2008), the brood parasitism behavior of the cuckoo species in the cuckoo search (Yang and Deb 2009), the echolocation behavior of microbats in the bat algorithm

(Yang 2010b), the pollination process of flowering plants in the flower pollination algorithm (FPA) (Yang 2012), the herding behavior of krill in the krill herd algorithm (KH) (Gandomi and Alavi 2012) and the behavior of bees in the honey bee algorithm (HBA) (Karaboga 2005), the virtual bee algorithm (VBA) (Yang 2005) and the artificial bee colony (ABC) (Karaboga and Basturk 2008). The details of swarm intelligence algorithms are given by Parpinelli and Lopes (2011). In this section, PSO, FPA and BA are briefly explained.

2.4.1. *Particle swarm optimization*

PSO is a simplified algorithm since it does not contain the as mutation and crossover features. In PSO, the solutions of design variables are defined as position vectors (x_i^t for i=1, 2, 3, ...,N) and the design variables are mentioned as particles. Each particle has a velocity (v_i^t for i=1, 2, 3, ...,N) for a swarm with N particles. The ith position vector is calculated according to the following equation:

$$x_i^{t+1} = x_i^t + v_i^{t+1} \tag{2.4}$$

where the symbol t in the upper script represents the iteration number of the iterative process. For the initial solutions (t=0), the initial position vectors can be randomly generated, while the initial velocity is zero. The velocity is calculated according to equation [4.5].

$$v_i^{t+1} = \theta(t)v_i^t + \alpha(\text{rand}(1))(g^* - x_i^t) + \beta(\text{rand}(1))(x^* - x_i^t) \tag{2.5}$$

where α and β are learning parameters that can typically be taken as 2, and g^* and x^t are the best of global and local solutions, respectively. The inertia function (t) is an additional factor that does not exist in the classical form of PSO. It may be needed to control the weight of the velocity and a constant value can also be given instead of a function of the iteration number. g^* is the best solution of all of the times, but x^* is the best one of an iteration.

Example: Basic application for iterative stages of PSO

Assume that x ℝ, where the objective is to minimize the function:

$$f(x_i) = |x_i + 1| \text{ for } i=1, 2, 3, ..., N \tag{2.6}$$

The initial solutions (t=0) and the first four iteration solutions (t=1, 2, 3, 4) are given in Table 2.1 by taking $\alpha=\beta=2$ and (t)=1. The number of particles (N) is equal to 5. It is seen that the best objective function for initial solutions is 0.89. After four iterations, the best value is 0.05.

Iteration (t)	Symbol	i=1	i=2	i=3	i=4	i=5
0	v_i^t	0.00	0.00	0.00	0.00	0.00
	x_i^t	2.12	12.54	5.32	-8.11	-1.89
	$f(x_i)$	3.12	13.54	6.32	7.11	0.89
	g^*	-1.89				
	x^*	-1.89				
1	rand(1)	0.54	0.34	0.80	0.82	0.24
	v_i^t	-8.05	-16.80	-16.79	26.75	1.97
	x_i^t	-5.93	-4.26	-11.47	18.64	0.08
	$f(x_i)$	4.93	3.26	10.47	19.64	1.08
	g^*	-1.89				
	x^*	0.08				
2	rand(1)	0.13	0.29	0.96	0.97	0.21
	v_i^t	-5.44	-11.94	27.43	-45.93	1.90
	x_i^t	-11.38	-16.19	15.96	-27.28	1.98
	$f(x_i)$	10.38	15.19	16.96	26.28	2.98
	g^*	-1.89				
	x^*	1.98				
3	rand(1)	0.95	0.18	0.69	0.20	0.36
	v_i^t	38.19	-0.24	-16.42	-24.54	-0.93
	x_i^t	26.81	-16.43	-0.46	-51.83	1.05
	$f(x_i)$	27.81	15.43	0.54	50.83	2.05
	g^*	-0.46				
	x^*	-0.46				
4	rand(1)	0.05	0.90	0.76	0.68	0.28
	v_i^t	33.09	58.91	-15.02	116.93	-2.11
	x_i^t	59.90	42.48	-15.48	65.10	-1.05
	$f(x_i)$	60.90	43.48	14.48	66.10	0.05
	g^*	-1.05				
	x^*	-1.05				

Table 2.1. *Iterative stages of PSO example*

2.4.2. The flower pollination algorithm

The flower pollination algorithm developed by Yang (2012) is a nature-inspired metaheuristic algorithm formulating the pollination process of flowering plants. The types of flower pollination and flower constancy are effective in the development of the optimization processes called global and local pollination.

Pollination is the reproduction process of flowering plants. Flower constancy is a specialized flower–pollinator partnership. In this partnership, a specific flower can only attract specific species as pollinators. Thus, the flower constancy can be used as an increment step in the use of similarity or difference of two flowers. The flower constancy is used in both global and local pollination.

Global pollination uses the properties of two types of pollinations called biotic and cross-pollination. In biotic pollination, the pollinators are responsible for the pollen transfer process. Generally, 90% of flowering plants reproduce by using biotic pollinators. The pollinators may be insects, bees or other animals. Thus, the pollen transfer distance is up to the pollinator, and it is a random process. For that reason, the movement of pollinators obeys the rules of Lévy flights. Because of the long distance of pollen transfer, this type is an imitation of global pollination.

The pollination can be divided into two types by considering the flower involved. These types are cross-pollination and self-pollination. In global pollination, cross-pollination is the inspiration; it is a pollination type involving pollen from different flowers.

Global pollination can be written in a mathematical form as follows:

$$x_i^{t+1} = x_i^t + L(x_i^t - g^*),$$ [2.7]

where L is a Lévy distribution, g^* is the best existing solution of the design variable (in the mean of the objective function), x_i^t is the existing solution of the design variables for ith iteration, and x_i^{t+1} is the newly generated ith solution for (t+1)th iteration. This generation is done for a group of flowers or plants which represent a set of design variables.

Local pollination uses self-pollination as inspiration. The self-pollination process is a pollination process involving a single type of flower or plant. Also, biotic pollination is used in the development of local pollination. Abiotic pollination is the self-fertilization of flowers without a pollinator.

The mathematical form of local pollination uses a linear distribution (ε) and two randomly chosen flowers (j and k). The new solution is generated in equation [2.8]

$$x_i^{t+1} = x_i^t + \varepsilon(x_j^t - x_k^t) \qquad [2.8]$$

The classical form of FPA is proposed for single-objective problems, and a single pollen gamete is considered for simplification (Yang 2012). For multi-objective problems, Yang *et al.* (2014b) developed FPA considering multiple pollen gametes. Kayabekir *et al.* (2018a) introduced the application of FPA in engineering.

2.4.3. *The bat algorithm*

Bats use echolocation to sense distance and search for prey. When homing into their prey, a special frequency tuning is used. In this frequency tuning, the frequency, loudness and pulse emission rates vary. Yang (2010a) developed the bat algorithm (BA) by idealizing the echolocation behavior of bats.

In BA, the location of a bat is represented by the position vector. If the number of bats is equal to n, the position vector is represented by d_i^t, where the subscript i represents the location of the ith bat for i=1, 2, ..., n and the superscript t represents the iteration number. The design variables of engineering problem are stored in the position vectors, and a position vector of the next iteration (t+1) is generated according to

$$d_i^{t+1} = d_i^t + v_i^{t+1}, \qquad [2.9]$$

where v_i^{t+1} is the velocity vector of the (t+1)th iteration. The velocity vector is generated by adjusting the frequency (f_i) and the current best location (d^*), as shown in equation [2.10]

$$v_i^{t+1} = v_i^t + (d_i^t - d^*)f_i \qquad [2.10]$$

The frequency is randomly chosen between the minimum (f_{min}) and the maximum (f_{max}) frequency to equation [2.11].

$$f_i = f_{min} + (f_{max} - f_{min})(\text{rand}(1))$$ [2.11]

After the global generation of the position vector of the next iteration, solutions may or may not be accepted due to a criterion test which considers the pulse rate and loudness. If the pulse (r_i) is smaller than a randomly generated number, local search is used according to equation [2.12].

$$d_i^{t+1} = d_i^t + \varepsilon A_i^t$$ [2.12]

The local search represents a random walk using a linear random distribution ε [-1,1] and the loudness (A_i). Bats have a varying pulse rate and loudness. For that reason, the A_i^{t+1} and r_i^{t+1} values of (t+1)th iterations are as follows:

$$A_i^{t+1} = \alpha A_i^t,$$ [2.13]

$$r_i^{t+1} = r_i^0[1 - \exp(-\gamma^t)],$$ [2.14]

where α and γ are constant values used to control the range and usage of the local search.

2.5. Other metaheuristic algorithms

2.5.1. *Simulated annealing*

Simulated annealing (SA) mimics the annealing process in a material production process involving the annealing of metals. In annealing, a metal cools until it freezes into a crystalline state with minimum energy and a larger crystal size. Thus, the defects in the metallic schedule controlling the temperature of the cooling rate are important.

Kirckpatrick *et al.* (1983) proposed SA, which is a search method using a Markov chain to prevent being trapped in a local optima. The Markov chain converges under appropriate conditions concerning its transition probability. Changes improving the objective function are accepted, but some changes which are not ideal are also kept.

In SA, the initial temperature (T_0) and initial guess (x^0) are chosen, but the initial temperature is critical. If it is too high ($T \to \infty$), all changes are accepted. If it is too low ($T \to 0$), all changes are rejected.

In the iterative process, the new locations representing the possible design variables are generated by using are the random generation:

$$x_i^{t+1} = x^t + randn,$$ [2.15]

where randn is a normally distributed pseudorandom number. This generation is done until the temperature (T) is higher than the final temperature (T_f) and the iteration number (t) is lower than the maximum iteration number (t_{max}):

$$T > T_f \text{ and } t < t_{max}.$$ [2.16]

The new solution (x^{t+1}) is accepted if it is better than the old one (x^t). If not, the new solution is accepted when the p value:

$$p = exp[-\Delta f/T] > r,$$ [2.17]

where r is a random number and Δf is the difference of the objective function of two last solutions ($f(x^{t+1})$) and $f(x^t)$:

$$\Delta f = f(x^{t+1}) - f(x^t).$$ [2.18]

The temperature (T) is found according to a cooling schedule, which can be linear or geometric, as seen in equations [2.19] and [2.20], respectively:

$$T(t) = T_0 - \beta t$$ [2.19]

$$T(t) = T_0 \alpha^t$$ [2.20]

where β is the cooling rate and α is the cooling factor. By using equation [2.20], $T \to 0$ when $t \to \infty$. For that reason, the usage of the maximum iteration number is not necessary since the cooling process should be slow. α is proposed to be between 0.7 and 0.95 (Yang 2010b). Different formulations for the cooling schedule can be also used. All of the best solutions are stored, but the iterations continue, according to the accepted solution. A survey on simulated annealing was developed by Suman and Kumar (2006).

2.5.2. *Teaching–learning-based optimization*

Rao *et al.* (2011) developed teaching–learning-based optimization (TLBO) by formulating two phases of education in a class. These two phases are the teacher and learner phases.

A teacher aims to express knowledge to their students. In this phase, the teacher is the member of the class with the best knowledge. For that reason, the teacher phase is orientated on the usage of the best solution (g^*) as a teacher. The generation in the teacher phase is done according to equation [2.21] for a class with n population.

$$x_i^{t+1} = x_i^t + rand(1)(g^* - TF\ x_{ave}) \tag{2.21}$$

In equation [2.21], the average of all solutions,

$$x_{ave} = \frac{\sum_{i=1}^{n} x_i^t}{n} \tag{2.22}$$

is used with an algorithm parameter called the teaching factor (TF), which is a randomly generated integer number, and TF can only gain the values 1 or 2. This parameter controls the range of the convergence to the best solution. TF is not a user-defined parameter, and this feature is an advantage of TLBO. Different from the other algorithms, the only user-defined values are the population of the class and the maximum iteration number. A parameter or rate is also not needed to control the type of phase, because, in real life, the teacher and learner phases are sequential stages. After the education process of the teacher, a group study of the learners is done in the learner (student) phase. The learner phase aims to increase the level of knowledge of all of the students, and the next teacher may be one of these students. The equation of the learner phase is as follows:

$$x_i^{t+1} = \begin{cases} x_i^t + rand(1)(x_j^t - x_k^t) \ if \ f(x_j^t) < f(x_k^t) \\ x_i^t + rand(1)(x_k^t - x_j^t) \ if \ f(x_j^t) > f(x_k^t) \end{cases} \tag{2.23}$$

where $f(x_k^t)$ and $f(x_j^t)$ are the objective functions of two randomly chosen solutions, kth and jth, respectively. In equation [2.23], the formulations are written for the minimization of the objective function. If the objective is to maximize the objective function, the inequalities in equation [2.23] must be vice versa. The learner phase aims to provide an effective convergence of all of the solutions of the population. The pseudocode of the TLBO optimization is given in Figure 2.2.

```
Objective minimize f(x)
Define class population and ranges
Randomly generate the initial students
while ( t<Max number of iterations )
        (Teacher Phase)
        Calculate the mean of each design variable (x_ave)
        Identify the best student as a teacher (g*)
        Generate new solutions (Eq. (2.21))
        Accept the new solution if better
        (Learner Phase)
        Select any two solutions randomly [ j, k]
        Generate new solutions (Eq. (2.23))
        Accept the new solution if better
        t=t+1
end while
```

Figure 2.2. *Pseudocode of TLBO algorithm*

2.5.3. *Harmony search*

The Harmony Search (HS) algorithm was developed by Geem *et al.* (2001), according to observations of musical performances. In a performance, a musician plays random notes. Then, the musicians aim to gain attention and admiration from the audience. For that reason, notes must be improved. To find the best music that is liked by listeners, new notes can be played. Also, the best notes in the memory of the musician can be played. By using this logical process, HS can be generated by including the following general steps:

Step 1: a harmony memory (HM) is initialized.

Step 2: a new harmony is improvised from HM.

Step 3: if the new harmony is better than the worst one in HM, include the new one and exclude the worst one.

Step 4: continue Steps 2 and 4 until the stopping criteria are satisfied.

To prevent trapping to a local optimum solution, two parameters are presented, called the harmony memory considering rate (HMCR) and pitch adjusting rate (PAR). In the first version of HS (Geem *et al.* 2001), HMCR and PAR are the two parameters. The third parameter is the harmony memory size (HMS), which is the number of notes (it is equivalent to the population number in other algorithms).

HMCR can be assigned with values between 0 and 1, and it is the probability to use an existing harmony in HM. Also, the neighboring values of the chosen existing harmony are used with a probability called PAR. PAR is also a value between 0 and 1. For example, if HMCR is equal to 0.5, the probability to choose an existing solution is 50%. If PAR is equal to 0.2, the probability to choose the previous value of the chosen harmony is 10%, while the probability for the next value of the chosen harmony is 10%, too.

Lee and Geem (2004) proposed HS for structural optimization. In the proposal, a neighborhood index (ni) and an arbitrary distance bandwidth (bw) are used for discrete and continuous optimization problems, respectively. A new harmony (equation [2.24]), including possible design variables, is generated as follows if r_1 and r_2 are random numbers between 0 and 1. r_3 is a random number -1 and 1:

$$x^{t+1} = x_{min} + rand(1)(x_{max} - x_{min}) \; if \; HMCR > r_1$$
$$x^{t+1} = x^t_j \; if \; HMCR \leq r_1 \; and \; PAR > r_2$$
$$x^{t+1} = x^t_{(j+ni)} \; if \; HMCR \leq r_1 \; and \; PAR \leq r_2 \; \text{(for discrete variables)}$$
$$x^{t+1} = x^t_j + bw(r_3) \; if \; HMCR \leq r_1 \; and \; PAR \leq r_2 \; \text{(for continuous variables)}$$

[2.24]

where ni is a random integer number that can take a negative or positive value and j is the index of the randomly chosen existing value. Several surveys were developed for the applications of HS (Maniarres *et al.* 2015; Siddique and Adeli 2015).

2.5.4. *The Jaya algorithm*

Rao (2016) developed a simple algorithm without specific control parameters and named the developed algorithm with the Sanskrit word, Jaya, which means victory. The developer generated Jaya after the success of

TLBO because it is a parameterless algorithm. In addition, Jaya only considers one phase instead of two, comparing to TLBO.

Jaya uses the best (g^*) and the worst (g^w) solutions. The algorithm aims to get closer to the best solution, while convergence to the worst solution is avoided. The formulation is as follows:

$$x_i^{t+1} = x_i^t + r_1(g^* - x_i^t) - r_2(g^w - x_i^t) \tag{2.25}$$

By equation [2.25], the algorithm always aims to get closer to the best solution, and also aims to avoid failure by moving away from the worst solution. Thus, victory is achieved.

3

Application of Metaheuristic Algorithms to Structural Problems

In optimization, the best solution for a set of design variables is found. The best solutions are decided according to a function that is related to the goal of the optimization. This function is called an objective function f(x), and is generally minimized or maximized for the optimum solutions, according to the problem.

Engineering problems generally also contain design constraints, which originate from physical rules, design regulations and architectural issues.

Briefly, optimization is to find the best possible design of a set of design variables (x) including N variables;

$$x = (x_1, x_2, ..., x_N)^T,$$ [3.1]

minimizing or maximizing an objective function or objective functions for an M objective optimization problem;

$$f_j(x_i), x \in \mathbb{R}, (i = 1, 2, ..., N) \text{ and } (j = 1, 2, ..., M);$$ [3.2]

providing K equality design constraints

$$h_k(x) = 0, (k = 1, 2, ..., K)$$ [3.3]

and Q inequality design constraints

$$g_q(x) \leq 0, (q = 1, 2, ..., Q).$$ [3.4]

In structural engineering, the design constraints are highly related to the design variables in optimization problems. For that reason, metaheuristic algorithms are effective in the problems of structural engineering.

3.1. Objective function

In this chapter, several types of objectives used in the optimization of structural engineering problems are mentioned. These are weight, cost, response, effectiveness and CO_2 emissions in construction.

The application of metaheuristics is not limited to the design optimization of structures. The nonlinear analyses of statically undetermined structural systems can be also analyzed by using Total Potential Optimization using Metaheuristic Algorithms (TPO/MA) (Toklu 2004a). In this methodology, the total potential energy of the structure is taken as the objective function, which is minimized for the precise solution of the structural system, taken as the design variables. Details and examples of TPO/MA are presented in Chapter 9.

3.1.1. *Weight*

Minimization of the total weight of structural systems is one of the best-known objectives in structural engineering. There are several advantages to constructing a light building.

The first advantage is saving on the material. A lighter building may have structural members with smaller cross-sections. Thus, this goal of structural engineering is achieved.

Economy factor: The total cost of the structure decreases.

Ecology factor: Lower CO_2 emissions are produced because of the production and transfer of materials.

Architectural and aesthetics factors: The use of a smaller cross-section is effective in the visual and architectural concept of a structure.

Comfort factor: Smaller cross-sections are easy on production and spacious areas may be provided.

Stress factor: By reduction of the weight of structural elements, the self-weight acting as the dead load of the structure is also reduced. In this case, the internal forces become lower and the stresses on the structural elements are reduced. Thus, structural members become safe and the failure of members may be prevented.

Seismic factor: The weight of the structure is directly related to the vertical load acting on the structure. According to Newton's second law (equation [3.5]), the force is the multiple of the acceleration (a) and mass (m) of the rigid body. Since the mass is the weight (W) of the rigid body divided by gravity (g), the increase of the weight affects the force (F):

$$F = ma = \frac{W}{g} a \hspace{4cm} [3.5]$$

3.1.2. Cost

Building a civil structure with low costs is one of the primary goals of engineers. Generally, the cost is reduced by designing a light structure, because the cost of materials is generally priced with their weights or volumes. If the price index is volume, it is also related to the weight of the structural member, according to the density of the material.

Generally, weight is the frequently used objective function for structures constructed using a single material. Labor costs may not be included when the weight is taken as the objective function, but designs with a single material may not show a great difference in labor costs.

Weight cannot be the only parameter in the minimization of the cost of composite materials like reinforced concrete. In reinforced concrete (RC), the components are concrete and steel and they have different unit prices and different behaviors. Therefore the parameter to be minimized should be combined cost, instead of weight. Steel is much more expensive than concrete, and thus this must be taken into account in the optimization process. It should also be noted that concrete has practically zero tensile strength and so steel bars, also known as rebar (reinforcing bar), should be used wherever there are tensile stresses in the structure.

The cost of the concrete is priced according to the unit volume of the members. If the cost of the concrete per unit volume is shown by C_{con}, the total cost of the concrete of a structure member is shown as follows:

$$Cost_{con} = C_{con}(Al_m - A_s l_s) \, , \tag{3.6}$$

where A is the cross-sectional area of the member, A_s is the area of the rebar in a cross-section of the member, l_m is the length of the member and l_s is the length of the rebar.

The steel is priced according to the unit weight. If the density of steel is γ_s, the total cost of the steel rebar ($Cost_{rebar}$) is given in equation [3.7], where C_{rebar} is the cost of the steel per unit weight:

$$Cost_{rebar} = C_{rebar} \gamma_s A_s l_s \tag{3.7}$$

In equations [3.6] and [3.7], the formulations are given for a single type of rebar. In general, different reinforcements, such as longitudinal reinforcements and shear reinforcements (stirrups), are separately designed. The longitudinal reinforcements are stored in the tensile section with A_s area and l_s length, but mounting (constructive) rebar or doubly reinforced members are placed in the compressive section of the members with A_s' area and l_s' length. Web reinforcements are also assigned if the depth of the cross-section is big. The area of web reinforcements is A_{web}, and the length of the web reinforcements is l_{web}. The shear reinforcements (stirrups) are generally designed for a unit length, and the area (A_w) is given for a constant spacing, which may be different in critical sections, with length l_c, and normal sections, with length l_n. In equations [3.8] and [3.9], the spacings are shown with s_c and s_n for critical and normal sections, respectively. The length of stirrups is shown with l_w:

$$Cost_{con} = C_{con} \left(Al_m - A_s l_s - A_s' l_s' - A_{web} l_{web} - (\frac{A_w}{s_c} l_c + \frac{A_w}{s_n} l_n) l_w \right) \tag{3.8}$$

$$Cost_{rebar} = C_{rebar} \gamma_s \left(A_s l_s + A_s' l_s' + A_{web} l_{web} + (\frac{A_w}{s_c} l_c + \frac{A_w}{s_n} l_n) l_w \right) \tag{3.9}$$

The cost of formwork (C_{form}) and labor (C_{labor}) can be included in the unit cost of concrete (C_{con}) and steel (C_{rebar}) in the calculation of the total cost of the structural member (C_{cost}), as given in equation [3.10]. The unit costs may also be prices purely for materials and formwork, and labor costs may be added to the total cost, as given in equation [3.11]:

$$Cost = C_{con} + C_{rebar} \qquad [3.10]$$

$$Cost = C_{con} + C_{rebar} + C_{labor} + C_{form} \qquad [3.11]$$

RC members are only an example of the usage of cost as an objective function. All design examples, including other types of composites and a retrofit purpose, are evaluated by minimizing the total cost.

3.1.3. *Response*

In the optimization of structures, a solution of analysis of the structural systems can be used as an objective to minimize or maximize. There are several examples of this approach in structural engineering.

The best-known technique is to minimize the internal forces of a structural system. This objective is provided by finding the optimum design variables, which are generally the location of members. For example, the optimum design problem may be finding the best spans for a structural design.

In documented methods, single- (Kayabekir *et al.* 2017) and multi- (Bekdaş *et al.* 2017a) story multi-bay frames are optimized for the minimization of the maximum stress by finding the best suitable span lengths of the structural system. In these studies, teaching learning-based optimization (TLBO) is used. As shown in Figure 3.1, the two design variables are x_1 and x_2 distances for the three-story frame structure. The structure is subjected to a distributed load (q). The stress on the members may be tensile or compressive stress. For that reason, the objective is to minimize the amplitude of the maximum stress, which has a minimum ($\sigma_{i,min}$) and a maximum ($\sigma_{i,max}$) in a cross-section of the ith member. The objective is to minimize the absolute value of $\sigma_{i,min}$ or $\sigma_{i,max}$ of a group of n members for i = 1, 2, 3, ..., n. The stress of the ith member is formulized in equation [3.12], where N_i, M_i, A_i and w_i are the axial force, flexural moment, area and section modulus of the ith member, respectively.

$$\sigma_{i,\min,\max} = \frac{N_i}{A_i} \mp \frac{M_i}{W_i}, \ (i = 1,2,....n)$$ [3.12]

For the structural system given in Figure 3.1, the fixed total length of spans (L_{span}) is 15 m and the final optimum values of x_1 and x_2 are 4.95 m and 10.05 m, respectively, which is a symmetric design, since the geometry and loading of the system are symmetric. The critical members are the 19th and 21st members. The design constants used in the example are summarized in Table 3.1.

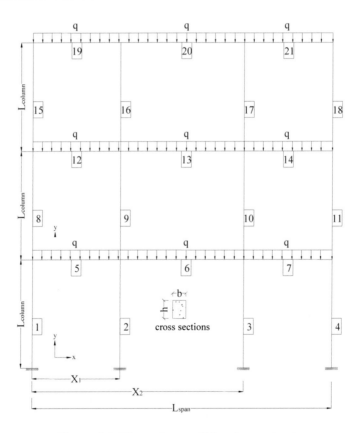

Figure 3.1. *Three-story multi-bay frame structure*

Definition	Symbol	Unit	Value
Total length of the span	L_{span}	m	15.0
Distributed load	q	kN/m^2	36.0
Elasticity modulus	E	MPa	200,000
Breadth of the elements	b	m	0.4
Height of the elements	h	m	0.4
Length of column	L_{column}	m	3

Table 3.1. *Design constants of a three-story frame example*

In an operation that is different from optimization, the structural systems can be analyzed by using a metaheuristic algorithm. In this area, the design with the minimum total potential energy is the solution. TPO/MA (Toklu 2004b) is also an example of using the response as an objective function.

In structural control applications, the objective is to reduce structural vibration. The reduction of the amplitude of structural vibration is one of the goals. In this goal, the effectiveness of the optimum control system is measured according to the reduction of a maximum structural response, compared to a structure without any control system. The response can be used as an objective of the structural control applications, but the effectiveness is a better index for the evaluation of the control system.

3.1.4. *Effectiveness*

The effectiveness of control systems is an important issue, and the performance of these systems is interrogated by the reduction of the ratio of the response of controlled and uncontrolled structures.

Tuned mass dampers (TMDs) are passive vibration absorber systems, and the properties of the mechanical components of the TMDs need to be optimized. Otherwise, we cannot expect effectiveness in reducing the responses of the structural system.

Generally, when finding the optimum values of parameters of a TMD, such as mass (or mass ratio), stiffness (or period) and damping coefficient (or damping ratio), the metaheuristic algorithm is adopted with either a time history (Bekdaş and Nigdeli 2011) or a frequency history (Nigdeli and

Bekdaş 2017b) analysis program. The design variables must be searched in a user-defined solution domain. The problem may be optimized for an objective function, which is the ratio of response (displacement, acceleration, shear force) of a TMD-controlled structure to an uncontrolled structure. The problem may have different objectives that are related to frequency domain responses or the stroke capacity of TMD. The stroke capacity may be also defined as a design constraint.

Similar to the TMD problem, the base isolation systems for the structures are optimized by using metaheuristic methods and effectiveness is the objective of the problem (Nigdeli *et al.* 2013). The details of TMD and base isolation problems will be given in Chapter 8.

3.1.5. *CO_2 emissions in construction*

In several studies about the optimization of RC structures, the minimization of CO_2 emissions is taken as a second objective of the optimization, in addition to the minimization of the total cost. These studies are as follows: Paya-Zaforteza *et al.* (2009) investigated the optimum design of RC frames by using simulated annealing (SA); Yepes *et al.* (2011) developed a variable neighborhood search threshold acceptance strategy for the optimum design of RC retaining walls; and The optimum design of RC footings employing a hybrid big bang–big crunch (BB-BC) algorithm was done by Camp and Assadallahi (2013).

CO_2 emissions result from the process of production and transfer of materials, such as concrete, reinforcements and formwork. In soil interacting structures, excavation and compacted backfill are also creators of CO_2 emissions.

3.2. The design constraints

The existence of design constraints is the main reason for using numerical optimization methods in structural engineering. The existence of design variables is the reason for the high nonlinearity of structural engineering problems.

The design constraints are generally oriented to the safety of the structural systems. Structural systems are designed according to mechanical rules and all physical interactions must be considered. For example, a civil structure may collapse because of the fracture of several members. If a member loses its load-bearing capacity, the health of the other members is less important. In this case, the serviceability of the structure is not ensured. For this reason, all members of a structural system must be sufficiently safe and the members with a higher capacity than required are not important. This situation makes optimization an important factor in structural design.

Ductility is one of the most important factors in structural design. In structural ductility, the structure must not lose the desired load-bearing capacity when several members are deformed, damaged or cracked. The columns of a structural system must be stronger than the beams. Deformation of the beams may not lead to a sudden collapse, but failure of the column would be a reason for collapse since the columns are the members carrying all the directed loads, which are directed from slabs to beams and beams to columns.

Another important ductility rule in structural engineering can be given from the design of RC beams. To prevent the fracture of brittle concrete in the compressive part of the section, before the yield of ductile steel, the maximum rebar in the tensile part of the section is limited.

The physical rules about the behavior of structures are coded in several design regulations. The structures designed must meet all the corresponding requirements. The structure must also meet the demand of individuals according to the intended purpose. For example, a small crack in an RC liquid tank may not be dangerous, but is important for liquid impermeability. The oscillation of bridges or tall buildings may also not be dangerous, but we must take into account the motion sickness that the movement may induce in individuals.

Architectural issues are also a constraint in structural engineering. Other aesthetic and comfort issues may also play an important role in design constraints. For example, the range of cross-sectional dimensions of a member or span distance in topology optimization may be determined as design constraints.

In this chapter, several types of design constraints are briefly described in the following sections, individually discussing stress, deformations, buckling, fatigue and design regulation rules.

3.2.1. *Stress*

In the analyses and design of structures, the determination of the stresses and deformations is involved. Before the stresses and deformations can be determined, the forces in the members are found.

As is very well known, the structural systems may be statically determined or statically indetermined. Statically determined systems can be solved by using equilibrium equations of statics. For statically indetermined systems, the equilibrium equations must be considered together with the geometry and material conditions of the problem. Several methods have been developed for solving statically indetermined structures, including the finite element method and TPO/MA (Toklu 2004c).

The definition of stress is the unit load on an area. There are several types of stresses, including normal and shear stresses.

Normal stresses are generally showed with the Greek letter σ and can occur because of axial loads and flexural moments. For example, the normal stress in a member under axial load (Figure 3.2) is calculated by dividing the axial load (N) by the cross-sectional area (A) of the perpendicular axis:

$$\sigma = \frac{N}{A} \qquad\qquad [3.13]$$

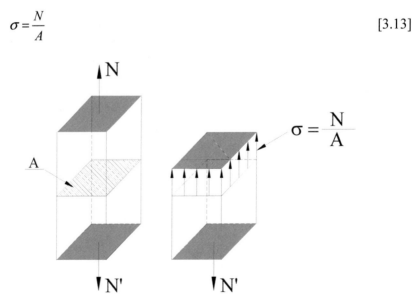

Figure 3.2. *Member with an axial load*

Under bending effect, changing stress occurs in the cross-section, as shown in Figure 3.3. The distance of the neutral surface and the distance of the cross-section are denoted as c and y, respectively. I is the moment of inertia of the cross-section. The equations for stress on the cross-section (σ_b) and the maximum absolute value of stress (σ_m), in the elastic range, are shown in equations [3.14] and [3.15], respectively.

$$\sigma_b = -\frac{My}{I}$$

[3.14]

$$\sigma_m = \frac{Mc}{I}$$

[3.15]

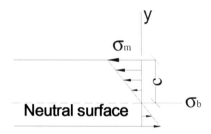

Figure 3.3. *Member under bending effect in elastic range*

The combined effect of the axial load and the flexural moments is shown in Figure 3.4. Frames are structural types where such effects can be seen. A typical example is given in section 3.1.3. The equation of the stress under the combined effect is as shown in equation [3.12] for a neutral axis at the middle of the cross-section.

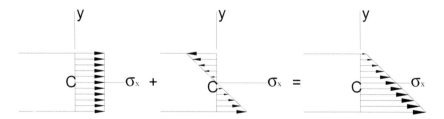

Figure 3.4. *Normal stress for the combined effect of axial load and flexural moment*

When transverse forces V and V', as shown in Figure 3.5, are applied to a member, a different type of stress called shearing stress comes into play. Shearing stress is shown with the Greek letter τ, and is given by equation [3.16]:

$$\tau(y) = \frac{VQ(y)}{b(y)I}$$

[3.16]

In equation [3.16], y is the coordinate of the point where the shearing stress is being calculated, $Q(y)$ is the first moment of the cross-sectional area above y with respect to the neutral axis of the section and b is the breadth of the section at y. The shear stress distribution on the transverse section of the rectangular beam is shown in Figure 3.6 with the maximum shearing stress (τ_{max}) at the level of the neutral axis, i.e. when $y = 0$.

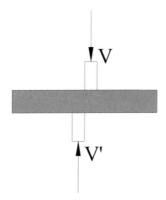

Figure 3.5. *A member with transverse loads*

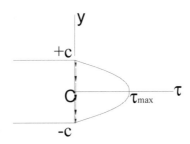

Figure 3.6. *Shear stress distribution on the transverse section of the rectangular section*

Details about the stress of mechanical systems can be found in the book by Beer *et al.* (2006). In this book, design considerations about stress are also given. These considerations are presented in the following topics:

– determination of the ultimate strength of a material (experimental test);

– allowable load and allowable stress (the safety factor of the design);

– selection of an appropriate safety factor;

– load and resistance factor design.

Stress limitations are a must in design optimization problems, including trusses, truss-like structures and frames. In addition to the stress, deformations, which have a relationship with stress, may also be considered as design variables.

3.2.2. *Deformations*

Deformation caused by the loads applied to a structure is another aspect in analysis and design. Large deformations may prevent the intended purpose of the structure from being achieved.

Analyses of deformations are useful for finding the stresses because of the relationship between stress and strain. The strain is the unit of deformation, which is generally denoted by the Greek letter ε, and is calculated using equation [3.17] for a bar with L length and a δ deformation under an axial load (N), as shown in Figure 3.7.

$$\varepsilon = \frac{\delta}{L} \qquad\qquad [3.17]$$

ε is the normal strain. According to Hook's law, the ratio of stress to strain is known as the elasticity modulus (E), as given in equation [3.18]:

$$E = \frac{\sigma}{\varepsilon} \qquad\qquad [3.18]$$

The directly linear behavior stress–strain relationship is seen up to the proportional limit, but practically, it can be used until the yielding of ductile materials and fracture of brittle materials.

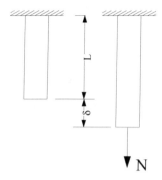

Figure 3.7. *Deformation of a bar under an axial load*

Shear stress (τ) does not have a direct effect on normal strain if the deformations remain small, but a small shear angle, defined as shearing strain (γ) occurs and the shear modulus (G) is used for the transfer of strain to stress for homogeneous isotropic materials.

The strain in one direction is also effected from a stress from another direction, according to Poisson's ratio (ν). The formulations of the normal and shearing strains are as follows:

$$\varepsilon_x = +\frac{\sigma_x}{E} - \frac{v\sigma_y}{E} - \frac{v\sigma_z}{E} \qquad [3.19]$$

$$\varepsilon_y = -\frac{v\sigma_x}{E} + \frac{\sigma_y}{E} - \frac{v\sigma_z}{E} \qquad [3.20]$$

$$\varepsilon_z = -\frac{v\sigma_x}{E} - \frac{v\sigma_y}{E} + \frac{\sigma_z}{E} \qquad [3.21]$$

$$\gamma_{xy} = \frac{\tau_{xy}}{G} \qquad [3.22]$$

$$\gamma_{yz} = \frac{\tau_{yz}}{G} \qquad [3.23]$$

$$\gamma_{zx} = \frac{\tau_{zx}}{G} \qquad [3.24]$$

where the subscripts for normal stresses and strains show the coordinates of the three directions (x, y and z), and the three planes (xy, yz and zx) are the subscripts of the shearing stresses and strains. The relationship between σ, E and v can be written as:

$$G = \frac{E}{2(1+v)}$$
[3.25]

3.2.3. Buckling

For the structural members with allowable stress, the design may be seen as suitable, but the members under a compressive axial load (Figure 3.8) will buckle instead of remaining straight, and then suddenly take a sharply curved position. Thus, the stability of the member is not ensured if the load is more than the critical load (N_{cr}).

Figure 3.8. *A buckled member*

Buckling is the reason for second-order effects where an additional flexural moment is formulized as

$$M = N(y)$$
[3.26]

where y is the deformation of the compressive member and N is the acting axial force. Because of the increase in the axial force, the deformations (y) approximate to infinite and the load-bearing capacity is lost through the buckling of the member.

Buckling must be used as a constraint in the design of members under axial loads, and is a critical factor for slender elements such as long columns.

3.2.4. *Fatigue*

If a structural member does not exceed the elastic limit of the material, it will return to its initial condition if the applied load is removed. If the loading is repeated a few thousands or millions of times, the rupture will occur at a stress that is lower than the design strength, and this phenomenon is called fatigue. This is a brittle failure even if the material is ductile. For this reason, fatigue is an important constraint for the optimum design of steel structures.

Fatigue is investigated by conducting experiments, and the results of these experiments are evaluated in design codes.

3.2.5. *Design regulation rules*

Civil engineering structures are designed according to the rules of the design regulations. These design regulations may show differences according to different countries. There are also several international organizations that develop design codes, including the American Concrete Institute (ACI), the Federal Emergency Management Agency (FEMA), the European Committee for Standardization (EUROCODE) and the American Institute of Steel Construction (AISC). Different design codes may include similar and distinctive design methods, approximations and empirical formulations of interpretations of experimental studies.

Some design codes for steel buildings are ANSI/AISC360-10 (2010) and EUROCODE 3 (2005). The general design concepts for steel structures are allowable strength design (ASD) and load and resistance factor design (LRFD). In ASD, a safety factor (SF) is used as a ratio of the ultimate stress of the structure (σ_u) and the allowable stress (σ_a) as given in equation [3.27]. The allowable stress is the yielding strength of steel:

$$SF = \frac{\sigma_u}{\sigma_a} \qquad [3.27]$$

In LRFD, the plastic capacity of steel is also considered and the allowed stress is the maximum strength of the steel. The general equation of LRFD is:

$$\varphi\sigma_u \geq \sum \gamma_i Q_i \qquad [3.28]$$

In equation [3.28], φ is the resistance factor and γ_i is the load factor for different types of loadings (Q_i) such as dead, live, wind, earthquake and snow loads.

The regulation rules about the design of steel structures include the following criteria for structural members.

i) Design criteria for typical tension members:

    – Nominal strength criteria:

        - yielding in full cross-sectional area;

        - rupture in punched cross-sectional area;

        - block shear failure;

        - slenderness requirements and stiffness criteria.

    – Special criteria for laterally supported steel beams:

        - formation of plastic hinges;

        - local buckling;

        - lateral torsional buckling.

ii) Design criteria for members under axial compressive forces:

    – crushing under compressive loads;

    – buckling, second-order effects.

iii) Design criteria for the member under flexural moments (M) and axial force (N):

  – M–N interaction curve (capacity);

  – N–δ deformation effect (demand).

iv) Details of steel structure projects:

  – details of conjunction points;

  – welding details;

  – mounting details.

The design of reinforced concrete structures is also subject to different regulations, such as Building Code Requirements for Structural Concrete (ACI318) (2008) and EUROCODE 2 (2004). Details on the optimization of RC structures is presented in Chapter 6. As a summary, the design criteria needed of RC structures are as follows.

– Principles of ductile structural design:

  - maximum rebar area for beams;

  - maximum axial force allowed;

  - stirrup design.

– Adherence criteria:

  - the ideal orientation of rebars;

  - overlapping distance;

  - anchorage distance.

– The load-bearing capacity of members:

  - M–N interaction curves;

  - moment magnification factor for second-order effects;

  - shear reinforcement design;

  - torsional effects;

  - punching (two-way shear) control.

4

Applications of Metaheuristic Algorithms in Structural Design

In this chapter, the generalities of structural optimization problems, sensibility analyses, reliability concepts of metaheuristic algorithms and general types of structural optimization are discussed. The numerical applications of several constrained simple structural engineering design optimization problems are presented. These problems include the vertical deflection minimization problem of an I-beam, the cost optimization of a tubular column under compressive load and the weight optimization of cantilever beams.

4.1. Generalities in structural design

The handling of multiple objectives is relatively easy when metaheuristic algorithms are used, compared to other optimization techniques. This can be done by combining all of the objectives into a single objective using appropriate weights. Some objectives, for instance the maximum displacement in a structure, do not necessitate a minimization or maximization process, but need to be within some predefined limitations. These objectives can be satisfied by introducing relevant constraints and/or by defining solution ranges in the procedure. The result obtained after the application of a multi-objective optimization will generally arrive at a Pareto front i.e. not a single solution, but to a multitude of solutions, among which the choice will be made by the users (Toklu 2005, Engau & Sigler, 2020).

4.2. Sensibility analyses and reliability

Metaheuristic algorithms rely on random search techniques, and the reliability of optimum results is always a question for designers. For this reason, an optimum design driven by the metaheuristic methodology must be statistically evaluated.

A sensibility analysis, which is a major way of evaluating the reliability of algorithms, is used to run the optimization process several times. In this case, the average outcome of the optimization objective can be measured. Moreover, an optimum design must not change according to multiple runs. The computational effort may show small differences, but the optimum design must be the same for the final results. In this case, the normal distribution of the final objective functions of different runs must have a small standard deviation value.

Another problem in structural optimization is getting trapped in the vicinity of a local optimum. For this reason, the standard deviation of the different runs of the optimization methodology is important. If a local result exists in a run of the methodology, the standard deviation value may increase greatly.

The convergence of the metaheuristic algorithm is also an index of comparison and reliability. The results must converge to the near optimum results quickly without trapping a local optimum, and the best current results must continue to improve to find a precise optimum solution. In addition, algorithm-specific parameters may play an important role in the convergence and prevent entrapping a local optimum.

In this section, basic structural engineering problems are optimized for several runs. The sensibility of the results is evaluated according to the algorithm parameters or population number.

4.3. Types of structural optimization

Structural optimization can be divided into three major types: sizing, shape and topology optimization. These optimization types address different aspects.

In sizing optimization, the optimum cross-section or dimensions of the elements in a structural system are searched, while the topology and shape of the elements are fixed. The counter or shape of elements is optimized in shape optimization for a fixed topology. In general, an optimal material layout of the structural system is constructed in topology or layout optimization (Kicinger *et al.* 2005).

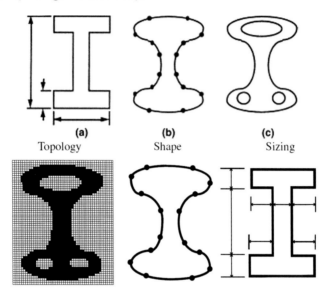

Figure 4.1. *Topology, shape and sizing optimization of an I-beam (Jakiela et al. 2000)*

The differences between the types of optimization are shown in Figure 4.1, for the optimum design of a cross-section of a beam. As seen in sizing optimization, the flange and web dimensions are optimized for a fixed-shaped I-beam, while the rectilinear dimensions and the parameterized shape remain constant. The shape is also optimized in shape optimization and the rectilinear dimensions are optimized as curved cross-sectional dimensions, but the topology is fixed. In topology optimization, new boundaries are defined in the material and holes can be created in the material.

The most popular structural engineering optimization exercise is truss structures. The difference between the types of truss structures is shown in

Figure 4.2, where the initial knowledge of the truss structure took an optimized form. In sizing optimization, the initial form is a classical truss system with a known topology and a cross-sectional shape. The thickness of the elements (or area) is optimized. Some of the elements may be assigned with zero areas, and these elements may be erased.

In shape optimization, a circular-holed initial form with constant topology is optimized. During topology optimization, the initial form is a rigid and fully filled body, which takes a truss-like form during the optimization process.

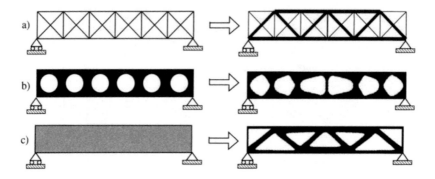

Figure 4.2. *Topology, shape and sizing optimization of structural optimization (Bendsøe and Sigmund 2004)*

4.4. Basic structural engineering applications

4.4.1. *Vertical deflection minimization problem of an I-beam*

The minimization of the vertical deflection of an optimum I-beam problem was first presented by Gold and Krishnamurty (1997). The optimization objective is to find the dimensions of the I-beam. The shape and topology are fixed. In this case, it is a sizing optimization problem. Four of the design variables are the flange width (b), the flange thickness (t_f), the beam height (h) and the web thickness (t_w), as shown in Figure 4.3.

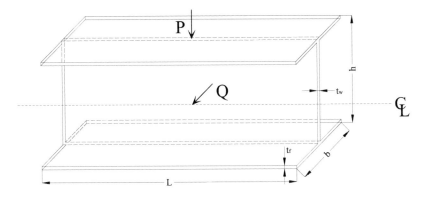

Figure 4.3. *I-beam problem*

The vertical deflection of the I-beam results from two design loads (P and Q) applied at midspan, vertically and horizontally. These loads are 600 kN and 50 kN for P and Q, respectively. The length of the beam (L) and the modulus of elasticity (E) are the design constants, which are taken as 200 cm and 20,000 kN/cm², respectively. The vertical deflection of the I-beam subjected to P load can be formulized as follows:

$$f(x) = \frac{PL^3}{48EI} \qquad\qquad [4.1]$$

The objective function of the vertical deflection minimization problem ($f(b,h,t_w,t_f)$) is given in equation [4.2]. In this equation, the numerical values of the design constants are applied to the parametric definition of the moment of inertia of the I-beam (I):

$$\text{Minimize } f(b,h,t_w,t_f) = \frac{5000}{\dfrac{t_w(h-2t_f)^3}{12} + \dfrac{bt_f^3}{6} + 2bt_f(\dfrac{h-t_f}{2})^2} \qquad [4.2]$$

The ranges of the design variables are given in equations [4.3]–[4.6] in cm:

$$10 \leq h \leq 80 \qquad\qquad\qquad [4.3]$$

$$10 \leq b \leq 50 \qquad\qquad\qquad [4.4]$$

$$0.9 \leq t_w \leq 5 \qquad\qquad\qquad [4.5]$$

$$0.9 \leq t_f \leq 5 \qquad\qquad\qquad [4.6]$$

The optimum design problem has two inequality constraints, denoted by g_1 and g_2. The cross-section of the I-beam must be less than 300 cm^2, which is the first design constraint formulated in equation [4.7]. Second, the bending stress of the beam must be less than 6 kN/cm^2, which is formulated in equation [4.8]:

$$g_1 = 2bt_f + t_w(h - 2t_f) \leq 300 \qquad\qquad [4.7]$$

$$g_2 = \frac{18000h}{t_w(h-2t_f)^3 + 2bt_w(4t_f^2 + 3h(h-2t_f))} + \frac{15000b}{t_w^3(h-2t_f) + 2t_wb^3} \leq 6 \qquad [4.8]$$

The Matlab (2010) codes of the methodology are presented in Appendix 1 for the Jaya algorithm (JA) (Rao 2016). The optimum results presented include 20 independent runs of the flower pollination algorithm (FPA) (Yang 2012; Nigdeli et al. 2016a) and the JA (Rao 2016). The maximum iteration number is 20,000. Moreover, the objective functions of the optimum results are presented for different population numbers, given in Tables 4.1 and 4.2.

The two methods are effective in finding accurate design variables with minimum vertical deflection. The JA is a parameter-free algorithm (other than the population number) with a single phase. The switch probability of the FPA was taken as 0.5. Tables 4.1 and 4.2 present the standard deviation of 20 runs, the iteration number of the optimization methodology and the generation number of design variables needed to find the best results.

Population number	2	3	5	10	15	20	30	50
Min. $f(b,h,t_w,t_f)$	0.0131	0.0131	0.0131	0.0131	0.0131	0.0131	0.0131	0.0131
Average $f(b,h,t_w,t_f)$	0.0602	0.027	0.0131	0.0131	0.0131	0.0131	0.0131	0.0131
Standard deviation	0.0487	0.034	2.4×10^{-4}	0	0	0	0	0
Iteration number	270	441	673	598	1,442	893	520	932
Generation number	540	1,323	3,365	5,980	2,163	1,786	1,560	4,660

Table 4.1. *Objective function of the I-beam problem for different population numbers (FPA)*

Population number	2	3	5	10	15	20	30	50
Min. $f(b,h,t_w,t_f)$	0.1632	0.0132	0.0131	0.0131	0.0131	0.0131	0.0131	0.0131
Average $f(b,h,t_w,t_f)$	0.1013	0.042	0.0131	0.0131	0.0131	0.0131	0.0131	0.0131
Standard deviation	0.1038	0.081	0	0	0	0	0	1.5×10^{-4}
Iteration number	256	446	1,174	652	489	633	731	711
Generation number	512	1,338	5,870	6,520	7,335	12,660	21,930	35,550

Table 4.2. *Objective function of the I-beam problem for different population numbers (JA)*

The optimum result cannot exactly be found when the population number of the JA is 2, as one of the two solutions is the best solution, while the other is the worst solution. In this case, the updated solution is the best or the worst, and a single part (+ or −) of the formulation of the JA works, but the trapping to a local optimum cannot be prevented. For this reason, the population number is an important index for the JA. The convergence plots for the case with a population of five solutions are shown in Figure 4.4. The JA has a faster convergence than the FPA.

The FPA and the JA are also compared with other methods used for the same problem. These methods are the adaptive response surface method (ARSM) (Wang 2003), the improved ARSM (Wang 2003) and the cuckoo search (CS) (Gandomi *et al.* 2013a). The optimum results are presented in Table 4.3.

	CS (Gandomi *et al.* 2013a)	ARSM (Wang 2003)	Improved ARSM (Wang 2003)	FPA (Yang 2012)	JA
H	80.00	80.00	79.99	80.00	80.00
B	50.00	37.05	48.42	50.00	50.00
t_w	0.9	1.71	0.9	0.9	0.9
t_f	2.3216715	2.31	2.40	2.3217922	2.3217922
$f(b,h,t_w,t_f)$	0.0130747	0.0157	0.131	0.0130741	0.0130741

Table 4.3. *Optimum results of the I-beam problem*

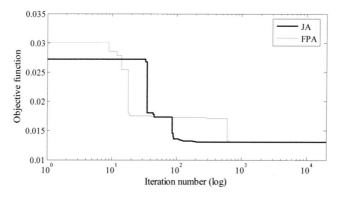

Figure 4.4. *Convergence graph for the I-beam problem. For a color version of this figure, see www.iste.co.uk/toklu/metaheuristics.zip*

4.4.2. *Cost optimization of the tubular column under compressive load*

The design optimization problem has two design variables, which are the average diameter (d) and the thickness of the tubular column (t) for the design of the cross-section with a constant shape. In this case, the problem is also a sizing optimization. This problem is shown in Figure 4.5, where $d_i=d-t/2$ and $d_0=d+t/2$. The design constants are given in Table 4.4. This problem was previously studied by Rao (1996), Hsu and Liu (2007) and Gandomi *et al.* (2013b) using the CS, and Nigdeli *et al.* (2016a) using the FPA.

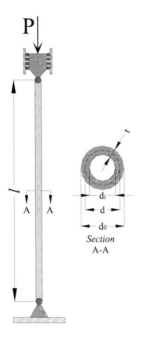

Figure 4.5. *Tubular column and A–A cross-section*

Symbol	Definition	Value
P	Axial force	2,500 kgf
σ_y	Yield stress	500 kgf/cm^2
E	Modulus of elasticity	0.85x10^6 kgf/cm^2
ρ	Density	0.0025 kgf/cm^3
l	Length of column	250 cm

Table 4.4. *Design constants of the tubular column*

The main objective of the tubular column problem is the minimization of the total material and the construction cost, as given in equation [4.9]:

$$f(d,t)=9.8dt+2d \tag{4.9}$$

The optimum design is subjected to two constraints that are related to the axial loading capacity and the buckling of the tubular column. As the first constraint (g_1), the compressive stress limit, which is the yield stress (σ_y), must not exceed in the design, as formulated in equation [4.10]:

$$g_1 = \frac{P}{\pi dt\sigma_y} - 1 \le 0 \tag{4.10}$$

Second, the Euler buckling load,

$$P_{kr} = \frac{\pi^2 EI}{l^2} \tag{4.11}$$

where I is the moment of inertia of the tubular column, must be less than the axial load on the column. It is formulated as a function of design variables, as given in equation [4.12]:

$$g_2 = \frac{8Pl^2}{\pi^3 Edt(d^2 + t^2)} - 1 \le 0 \tag{4.12}$$

In the following equations, the ranges of the design variables are formulated as four additional design constraints (g_3, g_4, g_5 and g_6):

$$g_3 = \frac{2}{d} - 1 \le 0 \tag{4.13}$$

$$g_4 = \frac{d}{14} - 1 \le 0 \tag{4.14}$$

$$g_5 = \frac{0.2}{t} - 1 \le 0 \tag{4.15}$$

$$g_6 = \frac{t}{0.8} - 1 \le 0 \tag{4.16}$$

The optimum results, including the six design constraints, are presented in Table 4.5.

	Hsu and Liu (2007)	Rao (1996)	CS (Gandomi *et al.* 2013b)	FPA (Yang 2012)
d	5.4507	5.44	5.45139	5.451160
t	0.292	0.293	0.29196	0.291965
g_1	-3.45×10^{-5}	-0.8579	0.0241	9.4343×10^{-7}
g_2	1.32×10^{-4}	0.0026	-0.1095	-4.249×10^{-7}
g_3	-0.6331	-0.8571	-0.6331	-0.6331
g_4	-0.6107	0	-0.6106	-0.6106
g_5	-0.3151	-0.7500	-0.3150	-0.3150
g_6	-0.6350	0	-0.6351	-0.6350
f(d, t)	26.4991	26.5323	26.53217	26.49948

Table 4.5. *Optimum results for the tubular column example*

For this example, the effect of the switch probability (p) of the FPA is tested. The results of 20 independent runs are given in Table 4.6. According to the results, when p ranges between 0 and 0.9, the accurate optimum results are found, but the number of iterations needed to reach the optimum value shows great differences. Al runs are effective in finding the same optimum value for p=0–0.9, but the last case (p=1) is not effective in finding the best solution and sensibility in finding the same solution for different runs. The number of iterations needed to reach the optimum value is 350 for p=0.6, which is a minimum number. Thus, the best way to adjust parameters is to give an almost equal probability for global and local searches. When only the local search is used (p=0), the computational effort needed is extremely high. If only the global search is used (p=1), the optimum result cannot be exactly found because of the weak convergence of the global search using a Lévy distribution. In this case, the importance of the parameter adjustment can be seen. The Matlab (2010) codes for the FPA are given in Appendix 2.

p	d	t	Min f(d,t)	Average f(d,t)	Standard deviation	Iteration number
0						14,820
0.1						1,187
0.2						727
0.3						527
0.4	5.4512	0.292	26.4995	26.4995	0	445
0.5						375
0.6						350
0.7						365
0.8						421
0.9						490
1.0	5.4507	0.292	26.5011	28.6465	2.38	1,194

Table 4.6. *Objective function and design variables of the tubular column problem for different switch probability numbers*

4.4.3. *Weight optimization of cantilever beams*

In this section, a cantilever beam problem is presented. The beam is rigidly supported from one end, while the other end is free. A vertical load is applied from the free end. The example has five different cross-sections to optimize the dimensions and has a square hollow cross-section, as shown in Figure 4.6.

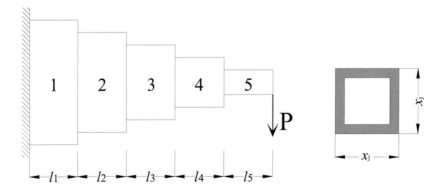

Figure 4.6. *The cantilever beam*

This example was presented by Fleury and Braibant (1986). The cantilever beam is divided into five steps with different sized cross-sections of the same shape. The thickness (t) of the hollow section is a design constant, which is taken as 2/3. For all steps of the beam, the design variables are x_j for j=1–5. In this case, the objective function, that is the total weight of the beam ($f(x_j)$), is as follows:

$$Minimize\ f(x_j) = 0.0624(x_1 + x_2 + x_3 + x_4 + x_5)$$ [4.17]

subjected to a design constraint:

$$g_1 = \frac{61}{x_1^3} + \frac{37}{x_2^3} + \frac{19}{x_3^3} + \frac{7}{x_4^3} + \frac{1}{x_5^3} - 1 \leq 0$$ [4.18]

ensuring the following range:

$$0.01 \leq x_j \leq 100$$ [4.19]

Methods	x_1	x_2	x_3	x_4	x_5	$f(x_j)$
MMA (Chickermane and Gea 1996)	6.0100	5.3000	4.4900	3.4900	2.1500	1.3400
GCA(I) (Chickermane and Gea 1996)	6.0100	5.3000	4.4900	3.4900	2.1500	1.3400
GCA(II) (Chickermane and Gea 1996)	6.0100	5.3000	4.4900	3.4900	2.1500	1.3400
CS (Gandomi et al. 2013a)	6.0089	5.3049	4.5023	3.5077	2.1504	1.33999
FPA (Nigdeli et al. 2016b	6.0202	5.3082	4.5042	3.4856	2.1557	1.33997
CONLIN (Chickermane and Gea 1996)	6.0100	5.3000	4.4900	3.4900	2.1500	NC

CONLIN: CONvex LINearization, MMA: method of moving asymptotes, GCA: generalized convex approximation

Table 4.7. *Optimum results of a cantilever beam (Example 1)*

The optimum results of the methods previously used for the example are presented in Table 4.7. The problem is also presented for 20 runs of the JA (Rao 2016) and teaching–learning-based optimization (TLBO) (Rao et al. 2011) by checking the different numbers of population in Tables 4.8 and 4.9, respectively. The maximum number of iterations is 20,000 for these two algorithms. As seen from the results, the JA needs less computational effort than TLBO, as the JA is a single-phase algorithm. TLBO is not affected by the population number when compared to the JA, but the multiple runs are not close to each other for a population of 2. The Matlab (2010) code for TLBO is given in Appendix 3.

Population number	2	3	5	10	15	20	30	50
Min. f(b,h,t_w,t_f)	3.470	1.429	1.340	1.340	1.340	1.340	1.340	1.340
Average f(b,h,t_w,t_f)	10.018	4.037	1.345	1.340	1.340	1.340	1.340	1.340
Standard deviation	3.179	2.185	0.02	0	0	0	0	0
Iteration number	269	9,313	17,927	16,263	15,119	15,311	19,231	19,598
Generation number	538	27,939	89,635	162,630	226,785	306,220	576,930	979,900

Table 4.8. Objective function of the cantilever beam problem for different population numbers (JA)

Population number	2	3	5	10	15	20	30	50
Min. f(b,h,t_w,t_f)	1.343	1.340	1.340	1.340	1.340	1.340	1.340	1.340
Average f(b,h,t_w,t_f)	1.464	1.343	1.340	1.340	1.340	1.340	1.340	1.340
Standard deviation	0.092	0.007	0	0	0	0	0	0
Iteration number	4,330	15,973	16,962	16,331	18,481	19,159	17,868	19,462
Generation number	8,660	47,919	84,810	163,310	277,215	383,180	536,040	973,100

Table 4.9. Objective function of the cantilever beam problem for different population numbers (TLBO)

4.5. Appendix 1[1]

```
% METAHEURISTICS FOR STRUCTURAL DESIGN AND ANALYSIS
% Chapter 5: Applications of Metaheuristic Algorithms in Structural
Design
% Section 5.4.1: Vertical deflection minimization problem of an I-beam

% clear memory
clear all

% sigma: allowable bending stress
% E: modulus of elasticity
% P: design load in vertical direction
% Q: design load in horizontal direction
% L: length of the beam
sigma=6; E=20000; P=600; Q=50; L=200;

% Algorithm parameters
% maxiter: maximum iteration number as stopping criteria
% pn: population number
maxiter=20000; pn=15;

% Ranges of design variables
% hmin: lower limit (minimum value) for height
% hmax: upper limit (maximum value) for height
% bmin: lower limit (minimum value) for breadth
% bmax: upper limit (maximum value) for breadth
% twmin: lower limit (minimum value) for web thickness
% twmax: upper limit (maximum value) for web thickness
% tfmin: lower limit (minimum value) for flange thickness
% tfmax: upper limit (maximum value) for flange thickness
hmin=10; hmax=80; bmin=10; bmax=50; twmin=0.9; twmax=5; tfmin=0.9;
tfmax=5;

% Generation of initial solution matrix
for i=1:pn
% Random generation of design variables
% h: height of the beam
% b: breadth of the beam
% tw: web thickness of the beam
% tf: flange thickness of the beam
% rand: a function in MATLAB that generates uniformly distributed
pseudorandom numbers
h=hmin+(hmax-hmin)*rand;
b=bmin+(bmax-bmin)*rand;
tw=twmin+(twmax-twmin)*rand;
tf=tfmin+(tfmax-tfmin)*rand;

% Objective function of the problem
Fx=5000/((((1/12)*tw*(h-2*tf)^3)+(((1/6)*b*tf^3)+(2*b*tf*((h-
tf)/2)^2))));
```

---

1 For a color version of the codes found in these Appendices, see www.iste.co.uk/toklu/
metaheuristics.zip.

```
% Inequality constraints of the problem
g1=2*b*tf+tw*(h-2*tf);
g2=(18000*h)/((tw*(h-2*tf)^3)+(2*b*tf*((4*tf^2)+(3*h*(h-
2*tf)))))+(15000*b)/(((h-2*tf)*tw^3)+(2*tw*b^3));

% OPT: Initial solution matrix
OPT(1,i)=h;
OPT(2,i)=b;
OPT(3,i)=tw;
OPT(4,i)=tf;
OPT(5,i)=Fx;
OPT(6,i)=g1;
OPT(7,i)=g2;

% Penalize the solutions that are not providing the inequalities
if OPT(6,i)>300
OPT(5,i)=10^6; % Penalized with a big value
end

if OPT(7,i)>6
OPT(5,i)=10^6; % Penalized with a big value
end

end

% Iterative process
for iter=1:maxiter

% Generation of new solutions according to rules of the JA
for i=1:pn

%Selection of the best existing solution and its design variables
[p,r]=min(OPT(5,:));
best1=OPT(1,r);
best2=OPT(2,r);
best3=OPT(3,r);
best4=OPT(4,r);

%Selection of the worst existing solution and its design variables
[k,t]=max(OPT(5,:));
worst1=OPT(1,t);
worst2=OPT(2,t);
worst3=OPT(3,t);
worst4=OPT(4,t);

%Generation of design variables according to JA rules (see equation
[3.25] in Chapter 3)
h=OPT(1,i)+rand*(best1-abs(OPT(1,i)))-rand*(worst1-abs(OPT(1,i)));
b=OPT(2,i)+rand*(best2-abs(OPT(2,i)))-rand*(worst2-abs(OPT(2,i)));
tw=OPT(3,i)+rand*(best3-abs(OPT(3,i)))-rand*(worst3-abs(OPT(3,i)));
tf=OPT(4,i)+rand*(best4-abs(OPT(4,i)))-rand*(worst4-abs(OPT(4,i)));

% Checking the upper and lower limits of generated values
if h>hmax
h=hmax;
end
if h<hmin
```

```
h=hmin;
end

if b>bmax
b=bmax;
end
if b<bmin
b=bmin;
end

if tw>twmax
tw=twmax;
end
if tw<twmin
tw=twmin;
end

if tf>tfmax
tf=tfmax;
end
if tf<tfmin
tf=tfmin;
end

% Objective function of the problem
Fx=5000/((((1/12)*tw*(h-2*tf)^3)+(((1/6)*b*tf^3)+(2*b*tf*((h-
tf)/2)^2))));

% Inequality constraints of the problem
g1=2*b*tf+tw*(h-2*tf);
g2=(18000*h)/((tw*(h-2*tf)^3)+(2*b*tf*((4*tf^2)+(3*h*(h-
2*tf))))))+(15000*b)/(((h-2*tf)*tw^3)+(2*tw*b^3));

%OPT1: New solution matrix
OPT1(1,i)=h;
OPT1(2,i)=b;
OPT1(3,i)=tw;
OPT1(4,i)=tf;
OPT1(5,i)=Fx;
OPT1(6,i)=g1;
OPT1(7,i)=g2;

% Penalize the solutions that are not providing the inequalities
if OPT1(6,i)>300
OPT1(5,i)=10^6;
end
if OPT1(7,i)>6
OPT1(5,i)=10^6;
end

end

% Compare the existing and newly developed solutions and save the
better one
for i=1:pn
if OPT(5,i)>OPT1(5,i)
OPT(:,i)=OPT1(:,i);
```

```
end
end

end
```

4.6. Appendix 2

```
% METAHEURISTICS FOR STRUCTURAL DESIGN AND ANALYSIS
% Chapter 5: Applications of Metaheuristic Algorithms in Structural
Design
% Section 5.4.2: Cost optimization of the tubular column under
compressive load

% clear memory
clear all

% sy: yield stress
% E: modulus of elasticity
% P: axial force
% % Li: length of a step
P=2500; E=0.85E6; sy=500; Li=250;

% Algorithm parameters
% maxiter: maximum iteration number as stopping criteria
% pn: population number
% ps: switch probability
maxiter=20000; pn=15; ps=0.5;

% Ranges of design variables (Constraints 3-6)
% dmin: lower limit (minimum value) for depth
% dmax: upper limit (maximum value) for depth
% tmin: lower limit (minimum value) for thickness
% tmax: upper limit (maximum value) for thickness
tmin=0.2; tmax=0.9; dmin=2; dmax=14;

% Generation of initial solution matrix
for i=1:pn
% Random generation of design variables
% d: depth of the column
% t: thickness of the column
% rand: a function in MATLAB that generates uniformly distributed
pseudorandom numbers
t=tmin+(tmax-tmin)*rand;
d=dmin+(dmax-dmin)*rand;

% Objective function of the problem
Fx=9.8*d*t+2*d;

% Inequality constraints of the problem
g1=P/(pi*d*t*sy)-1;
g2=8*P*(Li^2)/(pi^3*E*d*t*(d^2+t^2))-1;

% OPT: Initial solution matrix
OPT(1,i)=t;
OPT(2,i)=d;
```

```
OPT(3,i)=g1;
OPT(4,i)=g2;
OPT(5,i)=Fx;

% Penalize the solutions that are not providing the inequalities
if OPT(3,i)>0
OPT(5,i)=10^6; % Penalized with a big value
end

if OPT(4,i)>0
OPT(5,i)=10^6; % Penalized with a big value
end

end

% Iterative process
for iter=1:maxiter

%Generation of new solutions according to rules of the FPA
%Check for switch probability
if rand()<ps

%Global pollination
for i=1:pn
%Selection of the best existing solution and its design variables
[p,r]=min(OPT(5,:));
best1=OPT(1,r);
best2=OPT(2,r);

%Generation of a Lévy distribution
levy=(1/(2*pi()))*(rand()^-1.5)*exp(-1/(2*rand()));

%Generation of design variables according to global pollination (see
equation [3.7] in Chapter 3)
t=OPT(1,i)+levy*(best1-OPT(1,i));
d=OPT(2,i)+levy*(best2-OPT(2,i));

% Checking the upper and lower limits of generated values
if t>tmax
t=tmax;
end
if t<tmin
t=tmin;
end

if d>dmax
d=dmax;
end
if d<dmin
d=dmin;
end

% Objective function of the problem
Fx=9.8*d*t+2*d;

% Inequality constraints of the problem
g1=P/(pi*d*t*sy)-1;
```

```
g2=8*P*(Li^2)/(pi^3*E*d*t*(d^2+t^2))-1;

%OPT1: New solution matrix
OPT1(1,i)=t;
OPT1(2,i)=d;
OPT1(3,i)=g1;
OPT1(4,i)=g2;
OPT1(5,i)=Fx;

% Penalize the solutions that are not providing the inequalities
if OPT1(3,i)>0
OPT1(5,i)=10^6; % Penalized with a big value
end

if OPT1(4,i)>0
OPT1(5,i)=10^6; % Penalized with a big value
end

end

else

%Local pollination
for i=1:pn

%Random selection of two different solutions
xj=ceil(rand()*pn);
xk=ceil(rand()*pn);

%Generation of design variables according to local pollination (see
equation [3.8] in Chapter 3)
t=OPT(1,i)+rand()*(OPT(1,xj)-OPT(1,xk));
d=OPT(2,i)+rand()*(OPT(2,xj)-OPT(2,xk));

% Checking the upper and lower limits of generated values
if t>tmax
t=tmax;
end
if t<tmin
t=tmin;
end

if d>dmax
d=dmax;
end
if d<dmin
d=dmin;
end

% Objective function of the problem
Fx=9.8*d*t+2*d;

% Inequality constraints of the problem
g1=P/(pi*d*t*sy)-1;
g2=8*P*(Li^2)/(pi^3*E*d*t*(d^2+t^2))-1;

%OPT1: New solution matrix
```

```
OPT1(1,i)=t;
OPT1(2,i)=d;
OPT1(3,i)=g1;
OPT1(4,i)=g2;
OPT1(5,i)=Fx;

% Penalize the solutions that are not providing the inequalities
if OPT1(3,i)>0
OPT1(5,i)=10^6; % Penalized with a big value
end

if OPT1(4,i)>0
OPT1(5,i)=10^6; % Penalized with a big value
end

end

end

% Compare the existing and newly developed solutions and save the
better one
for i=1:pn
if OPT(5,i)>OPT1(5,i)
OPT(:,i)=OPT1(:,i);
end
end

end
```

4.7. Appendix 3

```
% METAHEURISTICS FOR STRUCTURAL DESIGN AND ANALYSIS
% Chapter 5: Applications of Metaheuristic Algorithms in Structural
Design
% Section 5.4.3: Weight optimization of cantilever beams

% clear memory
clear all
% Algorithm parameters
% maxiter: maximum iteration number as stopping criteria
% pn: population number
maxiter=20000; pn=15;

% Ranges of design variables
% xmin: lower limit (minimum value)
% xmax: upper limit (maximum value)
xmin=0.01; xmax=100;

% Generation of initial solution matrix
for i=1:pn
% Random generation of design variables
% xi: ith design variable
```

```matlab
% rand: a function in MATLAB that generates uniformly distributed
pseudorandom numbers
x1=xmin+(xmax-xmin)*rand;
x2=xmin+(xmax-xmin)*rand;
x3=xmin+(xmax-xmin)*rand;
x4=xmin+(xmax-xmin)*rand;
x5=xmin+(xmax-xmin)*rand;

% Objective function of the problem
Fx=0.0624*(x1+x2+x3+x4+x5);

% Inequality constraint of the problem
g1=((61/(x1^3))+(37/(x2^3))+(19/(x3^3))+(7/(x4^3))+(1/(x5^3)))-1;

% OPT: Initial solution matrix
OPT(1,i)=x1;
OPT(2,i)=x2;
OPT(3,i)=x3;
OPT(4,i)=x4;
OPT(5,i)=x5;
OPT(6,i)=Fx;
OPT(7,i)=g1;

% Penalize the solutions that are not providing the inequality
if OPT(7,i)>0
OPT(6,i)=10^6; % Penalized with a big value
end

end

% Iterative process
for iter=1:maxiter

% Generation of new solutions according to the rules of the TLBO

% Teacher Phase
for i=1:pn
%Selection of the best existing solution and its design variables
[p,r]=min(OPT(6,:));
best1=OPT(1,r);
best2=OPT(2,r);
best3=OPT(3,r);
best4=OPT(4,r);
best5=OPT(5,r);
%Generation of design variables according to the teacher phase rule
(see equation [3.21] in Chapter 3)
% TF: Teaching factor
% mean: a function in MATLAB that calculates the average or mean value
of array
TF=(round(1+rand));
x1=OPT(1,i)+rand*(best1-TF*mean(OPT(1,:)));
x2=OPT(2,i)+rand*(best2-TF*mean(OPT(2,:)));
x3=OPT(3,i)+rand*(best3-TF*mean(OPT(3,:)));
x4=OPT(4,i)+rand*(best4-TF*mean(OPT(4,:)));
x5=OPT(5,i)+rand*(best5-TF*mean(OPT(5,:)));

% Checking the upper and lower limits of generated values
```

```
if x1>xmax
x1=xmax;
end
if x1<xmin
x1=xmin;
end

if x2>xmax
x2=xmax;
end
if x2<xmin
x2=xmin;
end

if x3>xmax
x3=xmax;
end
if x3<xmin
x3=xmin;
end

if x4>xmax
x4=xmax;
end
if x4<xmin
x4=xmin;
end

if x5>xmax
x5=xmax;
end
if x5<xmin
x5=xmin;
end

% Objective function of the problem
Fx=0.0624*(x1+x2+x3+x4+x5);

% Inequality constraint of the problem
g1=((61/(x1^3))+(37/(x2^3))+(19/(x3^3))+(7/(x4^3))+(1/(x5^3)))-1;

% OPT1:New solution matrix
OPT1(1,i)=x1;
OPT1(2,i)=x2;
OPT1(3,i)=x3;
OPT1(4,i)=x4;
OPT1(5,i)=x5;
OPT1(6,i)=Fx;
OPT1(7,i)=g1;

% Penalize the solutions that are not providing the inequality
if OPT1(7,i)>0
OPT1(6,i)=10^6;
end

end
```

```
% Compare the existing and newly developed solutions and save the
better one
for i=1:pn
if OPT(6,i)>OPT1(6,i)
OPT(:,i)=OPT1(:,i);
end
end

% Learner Phase
for i=1:pn
%Generation of design variables according to the TLBO learner phase
rule (see equation [3.23] in Chapter 3)
% Random selection of two different solutions (learners)
xi=(ceil(rand*pn));
xj=(ceil(rand*pn));
while xi==xj
xi=(ceil(rand*pn));
xj=(ceil(rand*pn));
end

if OPT(6,xi)<OPT(6,xj)
x1=OPT(1,i)+rand*(OPT(1,xi)-OPT(1,xj));
x2=OPT(2,i)+rand*(OPT(2,xi)-OPT(2,xj));
x3=OPT(3,i)+rand*(OPT(3,xi)-OPT(3,xj));
x4=OPT(4,i)+rand*(OPT(4,xi)-OPT(4,xj));
x5=OPT(5,i)+rand*(OPT(5,xi)-OPT(5,xj));
else
x1=OPT(1,i)+rand()*(OPT(1,xj)-OPT(1,xi));
x2=OPT(2,i)+rand()*(OPT(2,xj)-OPT(2,xi));
x3=OPT(3,i)+rand()*(OPT(3,xj)-OPT(3,xi));
x4=OPT(4,i)+rand()*(OPT(4,xj)-OPT(4,xi));
x5=OPT(5,i)+rand()*(OPT(5,xj)-OPT(5,xi));
end

% Checking the upper and lower limits of generated values
if x1>xmax
x1=xmax;
end
if x1<xmin
x1=xmin;
end

if x2>xmax
x2=xmax;
end
if x2<xmin
x2=xmin;
end

if x3>xmax
x3=xmax;
end
if x3<xmin
x3=xmin;
end

if x4>xmax
```

```
x4=xmax;
end
if x4<xmin
x4=xmin;
end

if x5>xmax
x5=xmax;
end
if x5<xmin
x5=xmin;
end

% Objective function of the problem
Fx=0.0624*(x1+x2+x3+x4+x5);

% Inequality constraint of the problem
g1=((61/(x1^3))+(37/(x2^3))+(19/(x3^3))+(7/(x4^3))+(1/(x5^3)))-1;

% OPT1:New solution matrix
OPT1(1,i)=x1;
OPT1(2,i)=x2;
OPT1(3,i)=x3;
OPT1(4,i)=x4;
OPT1(5,i)=x5;
OPT1(6,i)=Fx;
OPT1(7,i)=g1;

% Penalize the solutions that are not providing the inequality
if OPT1(7,i)>0
OPT1(6,i)=10^6;
end
end

% Compare the existing and newly developed solutions and save the
better one
for i=1:pn
if OPT(6,i)>OPT1(6,i)
OPT(:,i)=OPT1(:,i);
end
end
end
```

Optimization of Truss-like Structures

Sizing optimization of pin-jointed structural systems (truss structures) is the most investigated structural engineering optimization problem. In this chapter, we first present a literature survey of the developed methods for the optimization of truss structures. Next, two basic, small benchmark truss structures are presented as numerical examples with Matlab (2010) code. One of these examples is a topology optimization and the other is a truss member sizing optimization. We then also present realistic truss structure optimization examples. Other truss-like structures, including tensegrity structures and cable nets, are also presented in this chapter.

5.1. The optimum design of truss structures

Truss structures consist of straight members that are connected at joints. These special structures have joints that can allow rotation and, in this case, truss members act as bars, which are structural members that only carry an axial force in either compression or tension. Since the joints of truss structures cannot allow fully free rotation, it is only a structural approximation in which bending effects are neglected because their contributions are small. These structures were popular, because of easy calculation, until the middle of the 20th century but truss-formed structural members are still elements that are commonly used for roofs, grids for suspension bridges and offshore steel structures (Krenk and Høgsberg 2013).

5.1.1. *Analyses of truss structures*

The number of degrees of freedom (n) of a truss structure with $N \geq 2$ nodes and $s \geq 0$ fixed nodal coordinate direction is calculated as follows:

$$n = dN - s \tag{5.1}$$

where d is the freedom of nodes, which is 2 for planar and 3 for spatial systems. Nodal displacements are equal to the global reduced coordinates of the structure (n). If equation [5.2] is provided for the truss system with m member of bars, small bars being overlapped by long bars can be prevented:

$$m \geq n \text{ and } m \leq \frac{N(N-1)}{2} \tag{5.2}$$

The normalized weight of the ith bar of an m bar structural system (λ_i) can be formulized in equation [5.3], where $\lambda_i \geq 0$. Equation [5.3] is only applicable for systems with the same material properties:

$$\lambda_i = \gamma L_i A_i, \, (A_i \in \mathbb{R}) \tag{5.3}$$

where A_i represents the cross-sectional area of the ith bar for $I = 1, \ldots, m$, γ is the density of all bars and L_i is the length of the ith bar.

The external forces of the truss structures are applied only at the nodal points and the forces are positioned in a vector; $P \in \mathbb{R}^n$ in the global reduced coordinates. If bending effects and gravity are neglected and deformations remain small, the equation of equilibrium can be written as follows:

$$K(A)u_A = P \tag{5.4}$$

where $u_A \in \mathbb{R}^n$ is the displacement vector of nodes in the global coordinates. The stiffness matrix of the truss structure is given as $K(A) \in \mathbb{R}^{n \times n}$, which is generated by merging the element stiffness matrices in the global coordinates; $K_i(A) \in \mathbb{R}^{2d \times 2d}$. For a spatial truss structure, $K_i(A)$ is as follows. The values of a, b and c are defined in equation [5.6]:

$$K_i(A) = EA_i \begin{bmatrix} a^2 & ab & ac & -a^2 & -ab & -ac \\ ab & b^2 & bc & -ab & -b^2 & -bc \\ ac & bc & c^2 & -ac & -bc & c^2 \\ -a^2 & -ab & -ac & a^2 & ab & ac \\ -ab & -b^2 & -bc & ab & b^2 & bc \\ -ac & -bc & c^2 & ac & bc & c^2 \end{bmatrix} \qquad [5.5]$$

$$a = \frac{L_{xi}}{L_i}, \ b = \frac{L_{yi}}{L_i} \ \text{and} \ c = \frac{L_{zi}}{L_i} \qquad [5.6]$$

In the stiffness matrix of the truss system, the corresponding rows and columns of fixed nodes (s) are eliminated, to find the stiffness matrix in global reduced coordinates. L_i denotes the full length of the ith bar, while L_{xi}, L_{yi} and L_{zi} are the dimensions of the bar in x, y and z global coordinates, respectively. The elasticity modulus is denoted by E, which is a material constant in the linear elastic design.

In the optimization of truss structures, two different types of design constraints are considered. The first includes the stress constraints $g_1(A)$, as formulized in the limitation of the normal stress (σ_i) of the ith bar with upper (σ^U) and lower (σ^L) limits [5.7]:

$$g_1(A): \sigma^L \leq \sigma_i \leq \sigma^U \ \ i = 1,.....,m \qquad [5.7]$$

The second constraint type is the limitation of displacements. If u^L and u^U, respectively, represent the lower and upper limits of displacement, the displacement constraint, $g_2(A)$, is as follows:

$$g_2(A): u^L \leq u_i \leq u^U \ \ i = 1,.....,N \qquad [5.8]$$

The σ^L represents the compression limitation ($\sigma^L \leq 0$), while the tensile limitation is denoted by σ^U ($\sigma^U \geq 0$).

The stress of the ith bar in global coordinates (σ_i^G) is calculated in equation [5.9], where u_i is the vector of nodal displacement of the ith bar:

$$\sigma_i^G = \frac{K_i(A)u_i}{A_i} \ \ i = 1,.....,m \qquad [5.9]$$

The upper and lower limits of the displacements are generally equal in absolute value and different in direction:

$$|u^L| = |u^U|; \, u^L \le 0, u^U > 0 \qquad\qquad\qquad [5.10]$$

5.1.2. *Review of the literature on the optimization of truss structures*

Metaheuristic-based methods have been widely proposed for the optimization of truss structures. Genetic algorithm (GA) and the variants of GA (hybrid or modified methods) are commonly used. Rajeev and Krishnamoorty (1992) used the GA, using discrete variables of the truss structure for practical optimum values, and proposed a penalty parameter depending on the violation of design constraints. The layout and sizing of trusses used in steel roofs are optimized by a GA-based methodology which was proposed by Koumousis and Georgiou in 1994. For the optimization of trusses, the GA is combined with min–max optimization (Coello and Christiansen 2000) and fuzzy formulations (Kelesoglu 2007). Sesok and Belevicius (2008) solved the topology optimization of trusses using a modified GA that uses a repairing genotype process instead of some constraints. An initial population strategy and self-adaptive member grouping were proposed by Togan and Daloğlu (2008) for trusses. A modified GA called species-conserving GA was developed and tested on truss structures by Li (2015).

Particle swarm optimization (PSO) was also used for the optimum design of truss structures (Schutte and Groenwold 2003; Perez and Behdinan 2007), as well as several variants that combined the particle swarm optimizer with the passive congregation, harmony search (HS) and ant colony optimization (ACO) (Li et al. 2007a; Kaveh and Talatahari 2009a). The other metaheuristic algorithms used in the optimization of truss structures are ant colony optimization (ACO) (Camp and Bichon 2004), teaching–learning-based optimization (TLBO) (Degertekin and Hayalioglu 2013; Camp and Farshchin 2014; Dede and Ayaz 2015), the artificial bee colony (ABC) (Sonmez 2011), the firefly algorithm (FA) (Miguel et al. 2013), cuckoo search (CS) (Gandomi et al. 2013a), the bat algorithm (BA) (Talathari and Kaveh 2015), Big Bang–Big Crunch (BB-BC) (Camp 2007; Kaveh and Talathari 2009b; Hasançebi and Kazemzadeh 2014), the mine blast algorithm (MBA) (Sadollah et al. 2012), the flower pollination algorithm

(FPA) (Bekdaş *et al.* 2015) and adaptive dimension search (Hasançebi and Azad 2015).

5.2. Numerical applications in the optimization of truss structures

This section includes four numerical applications of truss structures. The first example is a limited topology optimization and the angle of two bars concerning the horizontal axis is optimized for a symmetric 5-bar planar truss structure with fixed cross-sections. The second example is a 3-bar symmetric system with two cross-section design variables. These small examples are presented with Matlab (2010) code. In addition, two real-size structures, a 25-bar space structure and a 72-bar plane structure, are presented.

5.2.1. *A 5-bar truss structure optimization problem*

The numerical example, shown in Figure 5.1, is a symmetric structure, and the solution of the half system is adequate (Majid 1974). The main purpose of the optimization is to find the structure with the optimum values of angles θ_1 and θ_2, which are the angles between the horizontal axis and bars 1 and 2, respectively. The objective function ($f(\theta_1, \theta_2)$) formulized in equation [5.11] is the minimization of the total length of the half system, instead of the total weight, because the system has a fixed material density and cross-sectional area. The lengths of the bars (l_1, l_2, and l_3) can be calculated according to equations [5.12]–[5.14]:

$$Minimize\ f(\theta_1,\theta_2) = \sum_{i=1}^{3} l_i \qquad [5.11]$$

$$l_1 = \frac{l}{2\cos(\theta_1)} \qquad [5.12]$$

$$l_2 = \frac{l}{2\cos(\theta_2)} \qquad [5.13]$$

$$l_3 = \frac{l}{2\cos(\theta_1)\cos(\theta_2)}\sqrt{\cos^2(\theta_1)+\cos^2(\theta_2)-2\cos(\theta_1)\cos(\theta_2)2\cos(\theta_1-\theta_2)} \qquad [5.14]$$

The two inequality design constraints (g_1 and g_2), which are the limitations of vertical deflections of nodes 1 and 2 ($\Delta_1(\theta_1, \theta_2)$ and $\Delta_2(\theta_1, \theta_2)$) with the maximum allowed displacement (*MaxΔ*), are formulated in equations [5.15] and [5.16], respectively. The design variables are searched within a range with the minimum (θ_{min}) and maximum (θ_{max}) limits:

$$|\Delta_1(\theta_1, \theta_2)| \le Max\Delta \qquad\qquad [5.15]$$

$$|\Delta_2(\theta_1, \theta_2)| \le Max\Delta \qquad\qquad [5.16]$$

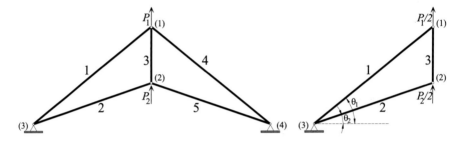

Figure 5.1. *The optimized system*

The cross-sectional area (A) and the elasticity modulus (E) of the bars are the design constants, and the values of all of the bars are the same. The structural system is subjected to two loads: P_1 and P_2. The distance between the two supports is symbolized by *l*.

If the displacement vector of nodes is generated as $\Delta = (\Delta_1, \Delta_1)^t$, the static equilibrium equation is written as $K\Delta = F$, where the element stiffness matrix and the load vector in the global coordinates are denoted by K and F, respectively. The global stiffness matrix is calculated as B^TkB, where B [5.17] and k [5.18] represent the transformation matrix of directional cosines and the stiffness matrix in the local coordinates, respectively:

$$B = \begin{bmatrix} \sin(\theta_1) & 0 \\ 0 & \sin(\theta_2) \\ 1 & -1 \end{bmatrix} \qquad\qquad [5.17]$$

$$k = \begin{bmatrix} \dfrac{EA}{l_1} & 0 & 0 \\[2ex] 0 & \dfrac{EA}{l_2} & 0 \\[2ex] 0 & 0 & \dfrac{EA}{l_3} \end{bmatrix} \qquad [5.18]$$

By calculating B^TkB with equations [5.17] and [5.18] and inserting it into the static equilibrium equation, the main analysis equation is obtained as follows:

$$EA \begin{bmatrix} \dfrac{\sin^2(\theta_1)}{l_1} + \dfrac{1}{2l_3} & -\dfrac{1}{2l_3} \\[2ex] -\dfrac{1}{2l_3} & \dfrac{\sin^2(\theta_2)}{l_2} + \dfrac{1}{2l_3} \end{bmatrix} \begin{bmatrix} \Delta_1 \\[1ex] \Delta_2 \end{bmatrix} = \begin{bmatrix} \dfrac{P_1}{2} \\[2ex] \dfrac{P_2}{2} \end{bmatrix} \qquad [5.19]$$

Table 5.1 presents the numerical values of the design constants and the ranges of the design variables. The problem is solved for different parameter cases of PSO, and the Matlab codes are presented in section 5.4, Appendix 1. The optimum values of the previously proposed algorithms FPA (Nigdeli *et al.* 2016a), GA (Li *et al.* 2007b) and CS (Gandomi *et al.* 2013b) are presented in Table 5.2.

Max Δ	5 mm
θ_{min}	0
θ_{max}	$\pi/3$ (rad)
A	100 mm^2
E	200,000 MPa
P_1	100 kN
P_2	50 kN
l	1,000 mm

Table 5.1. *Design constants of the 5-bar truss structure*

Method	θ_1(rad)	θ_2(rad)	$F(\theta_1, \theta_2)$
FPA	0.477634	0.477133	1125.87
GA	0.475784	0.472764	1125.98
CS	0.477459	0.477446	1125.92

Table 5.2. *Optimum results of the 5-bar truss structure*

In Table 5.3, the optimum results of the 5-bar example are presented for different inertia function values of PSO. The population number is 20 and both learning parameters (α and β) are taken as 2. The constant inertia function values are tested with 0.1 increments, and the cases with large inertia functions are not effective at finding the optimum value. The best inertia function value for the minimization of the objective function is 0.1, but the convergence (iteration number) and sensibility (standard deviation) of this case are not the best, given 30 independent runs of the optimization.

Inertia function	θ_1(rad)	θ_2(rad)	Min $f(\theta_1, \theta_2)$	Average $f(\theta_1, \theta_2)$	Standard deviation	Iteration number
0.1	0.4760	0.4760	1125.0797	1125.5912	0.2999	10,817
0.2	0.4773	0.4773	1125.8015	1125.8642	0.0238	4,705
0.3	0.4772	0.4772	1125.8736	1125.8901	0.0370	1,873
0.4	0.4764	0.4764	1125.3157	1125.9621	0.2405	3,822
0.5	0.4754	0.4795	1125.9369	1127.4423	2.3867	5,626
0.6	0.4752	0.4820	1126.6266	1136.2027	12.9234	1,186
0.7	0.4916	0.4770	1130.1410	1151.3288	12.7496	156
0.8	0.4794	0.4833	1128.2079	1159.5528	30.8627	1,246
0.9	0.4295	0.5324	1137.0161	1195.1854	62.7656	841
1.0	0.4696	0.4964	1129.7732	1194.2652	54.4259	3

Table 5.3. *Optimum results of the 5-bar truss structure for different inertia function values of PSO*

5.2.2. A 3-bar truss structure optimization problem

The sizing optimization of a 3-bar truss structure, shown in Figure 5.2 (Nowcki 1974), is presented in this section and the Matlab code using differential evolution (DE) is given in section 5.5. Appendix 2. Since the structural system is symmetric, the only design variables are the areas of the first and second bar (A_1 and A_2). The objective function ($f(A_1, A_2)$) is the minimization of the volume of the truss structure, as formularized in equation [5.20]:

$$Minimize\ f(A_1, A_2) = (2\sqrt{2}A_1 + A_2)l \qquad [5.20]$$

The optimization problem is subjected to the following design constraints (g_1, g_2 and g_3):

$$g_1 = \frac{\sqrt{2}A_1 + A_2}{\sqrt{2}A_1^2 + 2A_1A_2}P - \sigma \le 0 \qquad [5.21]$$

$$g_2 = \frac{A_2}{\sqrt{2}A_1^2 + 2A_1A_2}P - \sigma \le 0 \qquad [5.22]$$

$$g_3 = \frac{1}{A_1 + \sqrt{2}A_2}P - \sigma \le 0 \qquad [5.23]$$

The range of the cross-sectional areas is defined with the inequalities $0 \le A_1 \le 1$ and $0 \le A_2 \le 1$. Design constants length, load and stress limit are taken as $l = 100$ cm, $P = 2$ kN and $\sigma = 2$ kN/cm^2, respectively. The optimum results of the previously developed methods are presented in Table 5.4.

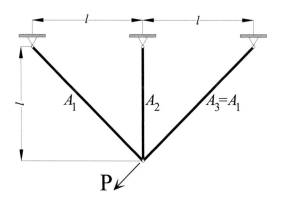

Figure 5.2. *Truss optimization problem*

st_max	Park *et al.* (2007)	Ray and Saini (2001)	Tsai (2005)	Yang and Gandomi (2012)	Gandomi *et al.* (2013a)	Nigdeli *et al.* (2016b)
A_1	0.78879	0.79500	0.788	0.78863	0.78867	0.78853
A_2	0.40794	0.39500	0.408	0.40838	0.40902	0.40866
$f(A_1, A_2)$	263.8965	264.3000	263.68	263.8962	263.9716	263.8958

Table 5.4. *Optimum results of the 3-bar truss structure (existing methods)*

F	A_1	A_2	Min $f(A_1, A_2)$	Average $f(A_1, A_2)$	Standard deviation	Iteration number
0.5	0.7887	0.4082	263.8958	263.8958	0	556
1.0	0.7887	0.4082	263.8958	263.9043	0.0238	154
1.5	0.7887	0.4082	263.8958	263.8959	$4*10^{-5}$	679
2	0.7887	0.4082	263.8958	264.0137	0.3174	207

Table 5.5. *Optimum results of the 3-bar truss structure (DE – different F)*

Population number	2	3	5	10	15	20	30	50
Min $f(A_1, A_2)$	283.01	267.2	270.81	264.25	263.95	263.90	263.89	263.89
Average $f(A_1, A_2)$	325.55	296.72	293.16	269.47	265.81	264.46	263.89	263.89
Standard deviation	33.71	17.49	19.13	7.27	2.18	0.64	0.006	0
Iteration number	4	4	4	15	29	43	218	477
Generation number	8	12	20	150	435	860	6,540	23,850

Table 5.6. *Optimum results of the 3-bar truss structure (DE – different population numbers)*

The DE algorithm is tested for different population numbers and control parameters (F). During optimization, the crossover constant (CR) is taken as 0.5. In Table 5.5, the optimum results are presented for 30 individuals of the population for different values of F. As seen from the results, the optimum results are similar to the other algorithms if F ranges between 0.5 and 1.5. For F = 1, the number of iterations to reach the optimum value is the lowest. For this reason, F is taken as 1 in the process of testing the population number, presented in Table 5.6. The results indicate that DE is not effective and robust for small population numbers. Indeed, Table 5.6 shows that for population numbers of less than, say, 20, the chances of getting trapped in the vicinity of a local optimum is very high, which means that standard deviation is also unacceptably high.

5.2.3. A 25-bar truss structure optimization problem

The spatial 25-bar truss structure is shown in Figure 5.3, and the elements are grouped as shown in Table 5.7. In this case, the number of design variables is 8 and the areas of the elements are optimized. The elasticity modulus of the elements is 10 Msi. The density is taken as 0.1 lb/in^3. The range of the cross-sectional areas of the elements is the same for all groups, between 0.01 and 3.4 in^2. The allowable displacement for each node in all coordinate directions is ±0.35 in, but the stress limits for compression and tension are different for all members, as presented in Table 5.7. The space truss is subjected to two independent loading conditions, as given in Table 5.8. The optimum results are presented in Table 5.9 for the different algorithms: ACO (Camp and Bichon 2004), heuristic particle swarm optimization (HPSO) (Li *et al*. 2007a), BB-BC (Camp 2007; Kaveh and Talathari 2009b), ABC (Sonmez 2011), TLBO (Camp and Farshchin 2014) and FPA (Bekdaş *et al*. 2015).

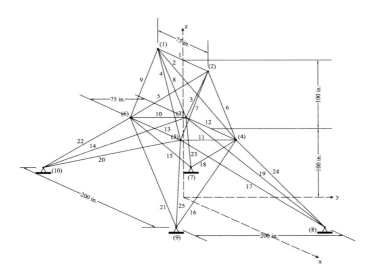

Figure 5.3. *A 25-bar space truss structure (Degertekin and Hayalioglu 2013)*

Element group	Members	Compression (ksi)	Tension (ksi)
1	1	35.092	35
2	2–5	11.590	35
3	6–9	17.305	35
4	10,11	35.092	35
5	12,13	35.092	35
6	14–17	6.759	35
7	18–21	6.959	35
8	22–25	11.082	35

Table 5.7. Member grouping and corresponding
stress limits for the 25-bar space truss structure

Case	Node	P_x (kips)	P_y (kips)	P_z (kips)
1	1	1.0	10.0	-5.0
	2	0.0	10.0	-5.0
	3	0.5	0.0	0.0
	6	0.0	0.0	0.0
2	1	0.0	20.0	-5.0
	2	0.0	-20.0	-5.0

Table 5.8. Multiple loading conditions for the 25-bar space truss

In Table 5.10, the variation trend of the optimized weight for the 25-bar space truss is presented for different combinations of population sizes (n) and the maximum number of iterations (t) of the FPA. We can see that the increasing population has a positive effect after the 1,000th iteration. If a small population size is used (n = 5), the FPA cannot find good results for small numbers of iterations.

Group	ACO (Camp and Bichon 2014)	HPSO (Li et al. 2007b)	BB-BC (Camp 2007)	BB-BC (Kaveh and Talathari 2009b)	ABC (Sonmez 2011)	TLBO (Camp and Farshchin 2014)	FPA (Bekdaş 2015)
1	0.0100	0.0100	0.0100	0.0100	0.0110	0.0100	0.0100
2	2.0000	1.9700	2.0920	1.9930	1.9790	1.9878	1.8308
3	2.9660	3.0160	2.9640	3.0560	3.0030	2.9914	3.1834
4	0.0100	0.0100	0.0100	0.0100	0.0100	0.0102	0.0100
5	0.0120	0.0100	0.0100	0.0100	0.0100	0.0100	0.0100
6	0.6890	0.6940	0.6890	0.6650	0.6900	0.6828	0.7017
7	1.6790	1.6810	1.6010	1.6420	1.6790	1.6775	1.7266
8	2.6680	2.6430	2.6860	2.6790	2.6520	2.6640	2.5713
Best weight (lb)	545.53	545.190	545.380	545.160	545.190	545.175	545.159
Average weight (lb)	545.34	-	545.780	545.660	-	545.483	545.730
Standard deviation on optimized weight (lb)	0.94	-	0.491	0.367	-	0.306	0.59
Number of structural analyses	16,500	125,000	20,566	12,500	500,000	12,199	8,149

Table 5.9. *Optimization results of the 25-bar truss space problem*

t/n	5	10	15	20	25	30
100	638.6077	604.8591	576.8526	585.4082	596.4658	588.3232
250	592.091	559.9247	560.0212	555.998	569.8168	567.2785
500	572.9288	559.9247	559.7946	554.8575	558.9415	556.6689
1000	553.2085	554.469	556.7446	553.0837	554.9815	556.3888
1500	553.2085	550.8044	556.7446	553.0837	547.3405	551.1369
2000	553.2085	550.8044	551.7404	551.1642	547.3405	548.4778

Table 5.10. *Sensitivity of the optimized weight of the 25-bar space truss structure*

5.2.4. A 72-bar space truss optimization example

The fourth problem is also a space truss, with 72 bars, as shown in Figure 5.4. The elasticity modulus and density of the material are equal to the values of the 25-bar space structure. The maximum allowed displacement of joints is ±0.25 in for all nodes, and the same stress limit in tension and compression is accepted for all elements of the structure (±25 ksi). The design variables defined as the cross-sectional areas of the elements must be between 0.1 and 3.0 in^2 and must carry two independent load cases, as shown in Table 5.11. The truss system is divided into 16 groups of design variables, as shown in Table 5.12, given with the optimum results of ACO (Camp and Bichon 2004), PSO (Perez and Behdinan 2007), BB-BC (Camp 2007; Kaveh and Talathari 2009b), TLBO (Camp and Farshchin 2014) and FPA (Bekdaş et al. 2015).

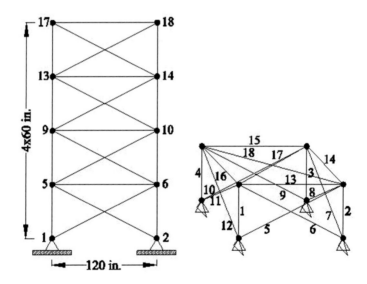

Figure 5.4. Schematic of the spatial 72-bar truss
structure (Camp and Farshchin 2014)

Case	Node	P_x (kips)	P_y (kips)	P_z (kips)
1	17–20	-5.0	-5.0	-5.0
2	17	5.0	5.0	-5.0

Table 5.11. Multiple load cases for the 72-bar truss

Element group	Members	ACO (Camp and Bichon 2004)	PSO (Perez and Behdinan 2007)	BB-BC (Camp 2007)	BB-BC (Kaveh and Talathari 2009b)	TLBO (Camp and Farshchin 2014)	FPA (Bekdaş et al. 2015)
1	1–4	1.9480	1.7430	1.8577	1.9042	1.8807	1.8758
2	5–12	0.5080	0.5180	0.5059	0.5162	0.5142	0.5160
3	13–16	0.1010	0.1000	0.1000	0.1000	0.1000	0.1000
4	17–18	0.1020	0.1000	0.1000	0.1000	0.1000	0.1000
5	19–22	1.3030	1.3080	1.2476	1.2582	1.2711	1.2993
6	23–30	0.5110	0.5190	0.5269	0.5035	0.5151	0.5246
7	31–34	0.1010	0.1000	0.1000	0.1000	0.1000	0.1001
8	35–36	0.1000	0.1000	0.1012	0.1000	0.1000	0.1000
9	37–40	0.5610	0.5140	0.5209	0.5178	0.5317	0.4971
10	41–48	0.4920	0.5460	0.5172	0.5214	0.5134	0.5089
11	49–52	0.1000	0.1000	0.1004	0.1000	0.1000	0.1000
12	53–54	0.1070	0.1090	0.1005	0.1007	0.1000	0.1000
13	55–58	0.1560	0.1610	0.1565	0.1566	0.1565	0.1575
14	59–66	0.5500	0.5090	0.5507	0.5421	0.5429	0.5329
15	67–70	0.3900	0.4970	0.3922	0.4132	0.4081	0.4089
16	71–72	0.5920	0.5620	0.5922	0.5756	0.5733	0.5731
Best weight (lb)		380.240	381.910	379.850	379.660	379.632	379.095
Average weight (lb)		383.160	-	382.080	381.850	379.759	379.534
Standard deviation on optimized weight (lb)		3.66	-	1.912	1.201	0.149	0.272
Number of structural analyses		18,500	8,000	19,679	13,200	21,542	9,029

Table 5.12. Optimization results and comparison
with the literature for the 72-bar truss problem

5.2.5. A 200-bar planar truss optimization example

Metaheuristic algorithms are also effective in sizing optimizations of large-sized truss structures. A 200-bar planar truss example is shown in Figure 5.5. The elasticity modulus and density of the elements of the structure are 30 Msi and 0.283 lb/in^3, respectively. Only stresses of the elements are constrained, and the equal limit stress in tension and compression is 25 ksi for all members. The range of all the cross-sectional areas of the elements is $0.1–20$ in^2. By grouping the elements, 29 sizing variables are available, as presented in Table 5.13, and the structure is subject to three independent load cases as follows:

i) +1 kips load in the X-direction at nodes 1, 6, 15, 20, 29, 34, 43, 48, 57, 62 and 71;

ii) -10 kips load in the Y-direction at nodes 1–6, 8, 10, 12, 14–20, 22, 24, 26, 28–34, 36, 38, 40, 42–48, 50, 52, 54, 56–62, 64, 66, 68, 70–75;

iii) superposition of the first and second loading conditions.

According to the three loading conditions, the optimum results are presented in Table 5.14 for the GA (Toğan and Daloğlu 2008), the hybrid method of HS, PSO and ACO (HPSACO) (Kaveh and Talathari 2009a), the TLBO (Degertekin and Hayalioglu 2013; Dede and Ayvaz 2015) and the FPA (Bekdaş et al. 2015).

Element group	Members	Element group	Members
1	1, 2, 3, 4	16	82, 83, 85, 86, 88, 89, 91, 92, 103, 104, 106, 107, 109, 110, 112, 113
2	5, 8, 11, 14, 17	17	115, 116, 117, 118
3	19, 20, 21, 22, 23, 24	18	119, 122, 125, 128, 131
4	18, 25, 56, 63, 94, 101, 132, 139, 170, 177	19	133, 134, 135, 136, 137, 138
5	26, 29, 32, 35, 38	20	140, 143, 146, 149, 152
6	6, 7, 9, 10, 12, 13, 15, 16, 27, 28, 30, 31, 33, 34, 36, 37	21	120, 121, 123, 124, 126, 127, 129, 130, 141, 142, 144, 145, 147, 148, 150, 151
7	39, 40, 41, 42	22	153, 154, 155, 156
8	43, 46, 49, 52, 55	23	157, 160, 163, 166, 169
9	57, 58, 59, 60, 61, 62	24	171, 172, 173, 174, 175, 176
10	64, 67, 70, 73, 76	25	178, 181, 184, 187, 190
11	44, 45, 47, 48, 50, 51, 53, 54, 65, 66, 68, 69, 71, 72, 74, 75	26	158, 159, 161, 162, 164, 165, 167, 168, 179, 180, 182, 183, 185, 186, 188, 189
12	77, 78, 79, 80	27	191, 192, 193, 194
13	81, 84, 87, 90, 93	28	195, 17, 198, 200
14	95, 96, 97, 98, 99, 100	29	196, 199
15	102, 105, 108, 111, 114		

Table 5.13. *Member grouping for the planar 200-bar truss structure*

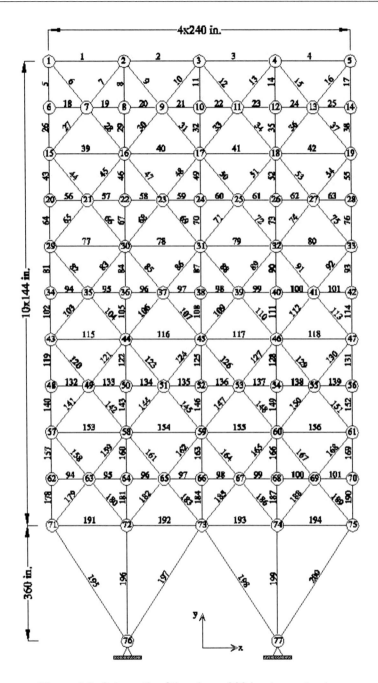

Figure 5.5. *Schematic of the planar 200-bar truss structure*

Element group	GA (Toğan and Daloğlu 2008)	HPSACO (Kaveh and Talathari 2009a)	TLBO (Degertekin and Hayalioglu 2013)	TLBO (Dede and Ayvaz 2015)	FPA (Bekdaş et al. 2015)
1	0.3469	0.1033	0.1460	0.1135	0.1425
2	1.0810	0.9184	0.9410	0.9484	0.9637
3	0.1000	0.1202	0.1000	0.1078	0.1005
4	0.1000	0.1009	0.1010	0.1000	0.1000
5	2.1421	1.8664	1.9410	1.9345	1.9514
6	0.3470	0.2826	0.2960	0.2889	0.2957
7	0.1000	0.1000	0.1000	0.2116	0.1156
8	3.5650	2.9683	3.1210	3.0903	3.1133
9	0.3470	0.1000	0.1000	0.1031	0.1006
10	4.8050	3.9456	4.1730	4.0903	4.1100
11	0.4400	0.3742	0.4010	0.4502	0.4165
12	0.4400	0.4501	0.1810	0.1007	0.1843
13	5.9520	4.9603	5.4230	5.4798	5.4567
14	0.3470	1.0738	0.1000	0.1011	0.1000
15	6.5720	5.9785	6.4220	6.4798	6.4559
16	0.9540	0.7863	0.5710	0.5329	0.5800
17	0.3470	0.7374	0.1560	0.1325	0.1547
18	8.5250	7.3809	7.9580	7.9445	8.0132
19	0.1000	0.6674	0.1000	0.1005	0.1000
20	9.3000	8.3000	8.9580	8.9444	9.0135
21	0.9540	1.1967	0.7200	0.7011	0.7391
22	1.7639	1.0000	0.4780	1.3777	0.7870
23	13.3006	10.8262	10.8970	11.2394	11.1795
24	0.3470	0.1000	0.1000	0.2287	0.1462
25	13.3006	11.6976	11.8970	12.2394	12.1799
26	2.1421	1.3880	1.0800	1.6849	1.3424
27	4.8050	4.9523	6.4620	4.9136	5.4844
28	9.3000	8.8000	10.7990	9.7190	10.1372
29	17.1740	14.6645	13.9220	15.0219	14.5262
Best weight (lb)	28544.0	25156.5	25488.2	25664.00	25521.81

Element group	GA (Toğan and Daloğlu 2008)	HPSACO (Kaveh and Talathari 2009a)	TLBO (Degertekin and Hayalioglu 2013)	TLBO (Dede and Ayvaz 2015)	FPA (Bekdaş et al. 2015)
Average weight (lb)	–	–	–	–	25543.51
Standard deviation on optimized weight (lb)	–	–	–	–	18.13
Number of structural analyses	-	9,875	28,059	-	10,685

Table 5.14. *Optimization results for the 200-bar truss optimization problem*

5.3. Tensegrity structures

Tensegrity structures are the combination of tensile elements (cables) and compressive elements (struts), forming a statically indeterminate structure in stable equilibrium. According to the Motro's description (1992), tensegrity structures are systems with rigidities resulting from a state of self-stressed equilibrium between cables under tension and compression elements, and are independent of all fields of action. The term "tensegrity" was first used by Fuller (1962) to describe the first tensegrity structure built by the sculptor Kennet Stenson.

The major aim in the design of tensegrity structures is form-finding, and several kinds of research have been conducted (Fuller 1962; Snelson 1965), including analytical nonlinear programming (Connelly and Terrell 1995), dynamic relaxation (Motro 1984) and kinematical (Tibert and Pellegrino 2003) methods, among others (Masic *et al.* 2005; Estrada *et al.* 2006; Zhang *et al.* 2006; Pagitz and Mirats Tur 2009). In general, recent studies about tensegrity structures' form-finding have proposed stochastic approaches, such as the Monte Carlo method (Li *et al.* 2010) and metaheuristic algorithms. The metaheuristic approaches employed include an evolutionary form-finding by Rieffel *et al.* (2009), GA (Xu and Luo 2010a; Koohestani 2012), simulated annealing (SA) (Xu and Luo 2010b), ACO (Chen *et al.* 2012) and differential evaluation (DE) (Do *et al.* 2016).

5.4. Appendix 1[1]

```
% METAHEURISTICS FOR STRUCTURAL DESIGN AND ANALYSIS
% Chapter 5: Optimization of Truss-like Structures
% Section 5.2.1: A 5-bar truss structure optimization problem

% clear memory
clear all

% E: modulus of elasticity
% A: cross-sectional area
% P1: design load
% P2: design load
% L: length
% Deltamax: maximum limit for Delta as inequality of the problem
E=200000; A=100; P1=100000; P2=50000; L=1000; Deltamax=5;

% Algorithm parameters
% maxiter: maximum iteration number as stopping criteria
% alfa: learning parameter
% beta: learning parameter
% w: inertia weight is a parameter that controls the previous velocity
maxiter=20000; pn=30; alfa=2; beta=2; w=0.1;

% Ranges of design variables
% Tetamin: lower limit (minimum value) for angle
% Tetamax: upper limit (maximum value) for angle
Tetamin=0; Tetamax=pi()/3;

% OPT: solution matrix
% OPT1: local solution matrix
% v: velocity vector
OPT=zeros(3,pn);
OPT1=zeros(3,pn);
v=zeros(2,pn);

% Generation of initial solution matrix
for i=1:pn
% Random generation of design variables
% T1: Teta1 angle
% T2: Teta2 angle
T1=Tetamin+rand*(Tetamax-Tetamin);
T2=Tetamin+rand*(Tetamax-Tetamin);

% L1: length of 1st member
% L2: length of 2nd member
% L3: length of 3rd member
L1=L/(2*cos(T1));
L2=L/(2*cos(T2));
L3=(sqrt(cos(T1)^2+cos(T2)^2-2*cos(T1)*cos(T2)*cos(T1-
T2))*L)/(2*cos(T1)*cos(T2));
```

1 For a color version of the codes found in these Appendices, see www.iste.co.uk/toklu/metaheuristics.zip.

```
% KG:Global stiffness matrix of the problem
KG=E*A*[sin(T1)^2/L1+1/(2*L3) -1/(2*L3);-1/(2*L3)
sin(T2)^2/L2+1/(2*L3)];

% D: Displacement vector
D=(KG^-1)*[0.5*P1; 0.5*P2];

% Objective function of the problem
Fx=L1+L2+L3;

% Deltamax: maximum limit for Delta as inequality of the problem
Deltamax=5;

% OPT:Initial solution matrix
OPT(1,i)=T1;
OPT(2,i)=T2;
OPT(3,i)=Fx;

% Penalize the solutions that are not providing the inequalities
if D(1,1)>Deltamax
OPT(3,i)=10^6; % Penalized with a big value
end

if D(2,1)>Deltamax
OPT(3,i)=10^6; % Penalized with a big value
end

end

% Iterative process
for iter=1:maxiter

% Generation of new solutions according to the rules of the DE
for i=1:pn
%Selection of the global best solution and its design variables
[p,r]=min(OPT(3,:));
best_global_1=OPT(1,r);
best_global_2=OPT(2,r);

%Selection of the local best solution and its design variables
[ce,be]=min(OPT1(3,:));
best_local_1=OPT1(1,be);
best_local_2=OPT1(2,be);

%Generation of design variables according to PSO rules (see equations
[3.4] and [3.5] in Chapter 3)
v(1,i)=w*v(1,i)+alfa*rand*(best_global_1-
OPT(1,i))+beta*rand*(best_local_1-OPT(1,i));
v(2,i)=w*v(2,i)+alfa*rand*(best_global_2-
OPT(2,i))+beta*rand*(best_local_2-OPT(2,i));
T1=OPT(1,i)+v(1,i);
T2=OPT(2,i)+v(2,i);

% Checking the upper and lower limits of generated values
if T1>Tetamax
T1=Tetamax;
end
```

```matlab
if T1<0
T1=Tetamax;
end

if T2>Tetamax
T2=Tetamax;
end
if T2<0
T2=Tetamax;
end

% L1: length of 1st member
% L2: length of 2nd member
% L3: length of 3rd member
L1=L/(2*cos(T1));
L2=L/(2*cos(T2));
L3=(sqrt(cos(T1)^2+cos(T2)^2-2*cos(T1)*cos(T2)*cos(T1-
T2))*L)/(2*cos(T1)*cos(T2));

% KG:Global stiffness matrix of the problem
KG=E*A*[sin(T1)^2/L1+1/(2*L3) -1/(2*L3);-1/(2*L3)
sin(T2)^2/L2+1/(2*L3)];

% D: Displacement vector
D=(KG^-1)*[0.5*P1; 0.5*P2];

% Objective function of the problem
Fx=L1+L2+L3;

% Deltamax: maximum limit for Delta as inequality of the problem
Deltamax=5;

%OPT1: New solution matrix
OPT1(1,i)=T1;
OPT1(2,i)=T2;
OPT1(3,i)=Fx;

% Penalize the solutions that are not providing the inequalities
if D(1,1)>Deltamax
OPT1(3,i)=10^6;
end
if D(2,1)>Deltamax
OPT1(3,i)=10^6;
end

end

% Compare the existing and newly developed solutions and save the
better one
for i=1:pn
if OPT(3,i)>OPT1(3,i)
OPT(:,i)=OPT1(:,i);
end
end

end
```

5.5. Appendix 2

```
% METAHEURISTICS FOR STRUCTURAL DESIGN AND ANALYSIS
% Chapter 5: Optimization of Truss-like Structures
% Section 5.2.2: A 3-bar truss structure optimization problem

% clear memory
clear all

% sigma: stress limit
% P: design load
% L: length
sigma=2; P=2; L=100;

% Algorithm parameters
% maxiter: maximum iteration number as stopping criteria
% pn: population number
% F: a control parameter that is a real constant number between 0 and 2
% CR: crossover constant which is assigned between 0 and 1
maxiter=20000; pn=30; F=1; CR=0.5;

% Ranges of design variables
% Amin: lower limit (minimum value) for cross-sectional areas
% Amax: upper limit (maximum value) for cross-sectional areas
Amin=0; Amax=1;

% Generation of initial solution matrix
for i=1:pn
% Random generation of design variables
% A1: cross-sectional areas for 1st and 3rd members
% A2: cross-sectional areas for 2nd member
A1=Amin+rand*(Amax-Amin);
A2=Amin+rand*(Amax-Amin);

% Objective function of the problem
Fx=(2*(2^0.5)*A1+A2)*L;

% Inequality constraints of the problem
g1=P*((2^0.5)*A1+A2)/((2^0.5)*A1^2+2*A1*A2)-sigma;
g2=P*(A2)/((2^0.5)*A1^2+2*A1*A2)-sigma;
g3=P/(A1+(2^0.5)*A2)-sigma;

% OPT: Initial solution matrix
OPT(1,i)=A1;
OPT(2,i)=A2;
OPT(3,i)=Fx;

% Penalize the solutions that are not providing the inequalities
if g1>0
OPT(3,i)=10^6; % Penalized with a big value
end
if g2>0
OPT(3,i)=10^6; % Penalized with a big value
end
if g3>0
OPT(3,i)=10^6; % Penalized with a big value
```

```
end

end

% Iterative process
for iter=1:maxiter

% Generation of new solutions according to the rules of the DE
for i=1:pn

% Selection of three random existing solutions
r1=(ceil(rand()*pn));
r2=(ceil(rand()*pn));
r3=(ceil(rand()*pn));

%Generation of design variables according to DE rules (see equation
[3.2] in Chapter 3)
v1=OPT(1,r1)+F*(OPT(1,r2)-OPT(1,r3));
v2=OPT(2,r1)+F*(OPT(2,r2)-OPT(2,r3));

%Accepting the mutated or existing solutions according to DE rules (see
equation [3.3] in Chapter 3)
r=(ceil(rand*pn));
if rand<=CR || i==r
A1=v1;
A2=v2;
else
A1=OPT(1,i);
A2=OPT(2,i);
end

% Checking the upper and lower limits of generated values
if A1>=1
A1=Amax;
end
if A1<=0
A1=Amin;
end

if A2>=1
A2=Amax;
end
if A2<=0
A2=Amin;
end

% Objective function of the problem
Fx=(2*(2^0.5)*A1+A2)*L;

% Inequality constraints of the problem
g1=P*((2^0.5)*A1+A2)/((2^0.5)*A1^2+2*A1*A2)-sigma;
g2=P*(A2)/((2^0.5)*A1^2+2*A1*A2)-sigma;
g3=P/(A1+(2^0.5)*A2)-sigma;

%OPT1: New solution matrix
OPT1(1,i)=A1;
OPT1(2,i)=A2;
```

```
OPT1(3,i)=Fx;

% Penalize the solutions that are not providing the inequalities
if g1>0
OPT1(3,i)=10^6;
end
if g2>0
OPT1(3,i)=10^6;
end
if g3>0
OPT1(3,i)=10^6;
end

end

% Compare the existing and newly developed solutions and save the
better one
for i=1:pn
if OPT(3,i)>OPT1(3,i)
OPT(:,i)=OPT1(:,i);
end
end

end
```

6

Optimization of Structures and Members

In the design of structures, the members are modeled according to the internal forces resulting from the loading conditions. Structures are not designed for a single loading case. The loads result from permanent and variable actions. The permanent actions contain the self-weight of the structure, including the structural and non-structural members. The non-structural members in a structure are all of the architectural components, such as exterior cladding, partitions and ceilings. If equipment or machinery always stays in the same position without changing the amount of loading (static load), these members are also permanent actions. The other loads, except permanent actions, are variable actions, and these loads may or may not exist. In this case, the internal forces may be positively or negatively affected by the loads and the section. For this reason, unfavorable loading conditions must be investigated. In addition, several variable actions are very short period loads, such as wind and earthquakes. Thus, the loads have a different factor of safety in different loading cases and all of the cases are necessary in the design to ensure the ultimate design limits considering stress–strain requirements. Because of the various loading cases, the optimum design of the structures is not found by mathematical methods and the problem is also nonlinear, since the design is effective in the analyses of structures. The various loading conditions and nonlinearity of the analyses according to the design are not the only difficulty in optimization. The design constraints must be taken into account, since a practical design is needed.

There are more challenges for composite and reinforced concrete (RC) structures. In the cost optimization of these structures, the usage of two different materials with different behaviors and the price are some of the obstacles. Generally, heuristic approaches are used in the optimization of structures and members. By generating design variables iteratively, all of the analyses and structural limit states can be easily controlled. Then, the design with the minimum cost, which does not violate design constraints, is the optimum solution.

6.1. Optimum design of RC beams

Because of the low tensile strength of concrete, the tensile stresses on a member are supported by reinforcement bars. In the calculations, the stresses on the tensile section only involve reinforcement forces. In Figure 6.1, the stresses on a rectangular cross-section are shown. The compressive stress block of concrete is a parabola. This parabola can be generally assumed to be a rectangular stress block with a smaller depth than the distance from the extreme compression fiber to the centroid of longitudinal tension reinforcement.

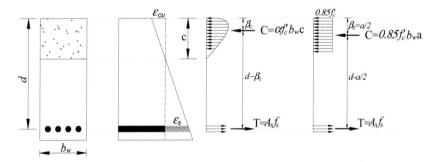

Figure 6.1. *The stresses on an RC beam cross-section*

In Figure 6.1, the symbols describe the following definitions:

b_w = web width or diameter of the circular section

d = distance from the extreme compression fiber to the centroid of longitudinal tension reinforcement

c = distance from the extreme compression fiber to the neutral axis

β_c = distance from the centroid of compressive stress block to the face of the compressive section

f_c' = specified compressive strength of concrete

f_s = calculated tensile stress in reinforcement at service loads

A_s = area of non-prestressed longitudinal tension reinforcement

a = depth of the equivalent rectangular stress block

ε_s = strain of steel in tensile section

ε_{cu} = strain of concrete in the face of the compressive section

C = resultant force of the compressive block

T = resultant force of tensile reinforcements

Since the forces shown in Figure 6.1 are in equilibrium (C=T), the distance c is written as follows:

$$c = \frac{A_s f_y}{\alpha f_c' b_w}$$ [6.1]

The depth of the equivalent stress block is defined as

$$a = \beta_1 c$$ [6.2]

In equation [6.2], β_1 is defined as a factor relating the depth of the equivalent rectangular compressive stress block to the neutral axis depth, and β_1 is defined according to the *fc'* values in ACI318: Building Code Requirements for Structural Concrete (ACI 318, 2005).

If f_c' is between 17 MPa and 28 MPa, β_1 is 0.85. If f_c' is above 28 MPa, β_1 linearly increases with 0.05 for each 7 MPa increase of compressive strength, but the smallest value of β_1 is 0.65.

If the equilibrium is written for a compressive block assumed to be rectangular, the distance a is given as follows:

$$a = \frac{A_s f_y}{0.85 f_c' b_w}$$ [6.3]

In this case, the moment capacity of the RC member can be formulated in two ways by using T or C forces, as shown in equations [6.4] and [6.5], respectively:

$$M = A_s f_s \left(d - \frac{a}{2} \right)$$
[6.4]

$$M = 0.85 f_c' b_w a \left(d - \frac{a}{2} \right)$$
[6.5]

The maximum strain of concrete (ε_c) in compression is generally taken to be 0.003. The failure is sudden if the concrete is crushed before steel. According to most of the specifications, the maximum reinforcement ratio (ρ_{max}) must not exceed 0.75 times the balanced reinforcement ratio so that a situation in which the concrete cracks and the steel yields at the same time is prevented. In this case, the maximum reinforcement ratio, which is the ratio of maximum reinforcement ($A_{s,max}$) to $b_w d$, is given as follows:

$$\rho_{max} = (0.75)(0.85)\beta_1 \frac{f_c'}{f_y} \left(\frac{600}{600 + f_y} \right)$$
[6.6]

f_y denotes the specified yield strength of the reinforcement. If the maximum ratio is not exceeded, a singly reinforced design (the only design with reinforcement bars in the tensile section and minimum constructive reinforcements in the opposite site) can be used and it is an economical case. In double reinforcement design with tensile and compressive reinforcements, the amount of tensile reinforcements shows an excessive increase to balance the compressive force of the reinforcements. Also, the market prices of steel are very expensive compared to those of concrete. In this case, the most economical and optimum design is to use singly reinforced sections and increase the cross-sectional dimensions if needed. The optimum results of an RC beam design generally aim to increase the cross-sectional dimensions. Of course, the ranges of the dimensions are limited to a maximum for architectural purposes and the optimization processes have to be governed to doubly reinforcement designs.

In ACI-318, the nominal flexural moment (M_n) is factored with ϕ to obtain the factored flexural moment, M_u. For tension controlled sections, ϕ is taken as 0.9. In ACI-318, the minimum reinforcement area must also be considered according to the two conditions formulated in equations [6.7] and [6.8]:

$$A_{s,min} \geq \frac{\sqrt{f_c'}}{4f_y} b_w d$$
[6.7]

$$A_{s,\min} \geq \frac{1.4}{f_y} b_w d \qquad [6.8]$$

The shear strength is provided by the sheer reinforcements or in other words, the stirrup. The nominal shear strength of concrete (V_c), shown in equation [6.9], is subtracted by the nominal shear force (V_u)

$$V_c = \frac{\sqrt{f_c'}}{6} b_w d \qquad [6.9]$$

Then, the factored shear force is calculated as follows:

$$V_u = \phi V_n \qquad [6.10]$$

The nominal shear strength provided by shear reinforcement is given in equation [6.11]:

$$V_s = \frac{A_v f_y d}{s} \qquad [6.11]$$

V_s must be less than $0.66\sqrt{f_c'} b_w d$ for brittle fracture. A_v denotes the area of shear reinforcement spacing s, and the minimum A_v is calculated in equation [6.12]. The maximum spacing between shear reinforcement ($s_{\max}$) must be taken as d/2, but $s_{\max}$ is d/4 if the V_s value is bigger than $0.33\sqrt{f_c'} b_w d$

$$\left(A_v\right)_{\min} = \frac{1}{3} \frac{b_w s}{f_y} \qquad [6.12]$$

The cost optimization of RC beams is one of the most popular applications that uses metaheuristic algorithms. GA-based approaches have been developed (Coello *et al.* 1997; Koumousis and Arsenis 1998), including continuous beams (Govindaraj and Ramasamy 2005) and T-shaped beams (Fedghouche and Tiliouine 2012). Also, GA is combined with simulated annealing (SA) in the optimum design of continuous RC beams, proposed by Leps and Sejnoha (2003). The optimum dimensions and detailed reinforcement designs of RC beams have also been carried out using metaheuristic approaches, such as the harmony search (HS) (Akin and Saka 2010; Bekdaş and Nigdeli 2012), teaching–learning-based optimization (Bekdaş and Nigdeli 2015) and bat algorithm (Bekdaş *et al.* 2014).

As a numerical example, RC beams under flexural effects are optimized by using three new generation metaheuristic algorithms. These algorithms are the flower pollination algorithm (FPA), teaching–learning-based optimization (TLBO) and Jaya algorithm (JA). The robustness of these algorithms is evaluated according to the best results, average results, standard deviation and the number of analyses needed to reach the best solution for 20 runs of the optimization process. The general methodology of the methods can be explained in the following five steps:

i) The objective flexural moment values, the design constants and the ranges of the design variables are defined.

ii) The initial solution matrix is generated. This matrix contains randomly generated candidate solutions of web width (b_w) and height (h) of the RC cross-sections.

iii) Then, the assignment of reinforcements is done according to the rules of ACI-318. In the analyses, the effective depth (d) of the beam is calculated considering clear cover, the orientation of longitudinal reinforcements and stirrups. If the reinforcement bars are not positioned in a single line because of the design code rules, the reinforcements are assigned in two lines and the effective depth (d) is recalculated. Also, when the singly reinforcement beam exceeds the maximum reinforcement area according to chosen cross-section dimensions, the doubly reinforcement beam design is used by generating the candidate compressive reinforcements. In addition, the minimum reinforcement conditions are checked. In the preliminary design, the depth of the beam (d) can be assumed by considering clear cover, minimum stirrup diameter and minimum longitudinal reinforcement, so that it can be decided whether the design is singly or doubly reinforced. After the reinforcements are assigned, the exact value of the depth and the required reinforcement area are updated. This step is repeated until the following three criteria are satisfied. By controlling these conditions, the unnecessary candidate solutions are eliminated and the possible negative effect of these solutions is prevented in the stage using metaheuristic algorithm rules. For example, a random existing solution is chosen and this solution may have a negative effect on convergence if it is too far from an optimum design. For the other design constraints defined according to ACI-318, the objective function can be penalized:

– The randomly assigned candidate reinforcements must not exceed the maximum reinforcement area for a singly reinforced design. In a doubly reinforced design, this condition is checked by considering the balanced condition according to concrete stress block and stresses on tensile and compressive reinforcements.

– The total area of reinforcements in candidate solutions must be between 1 and 1.05 times the required reinforcement area.

– The third criterion is only for the doubly reinforced design. The total area of the reinforcement of the compressive section must be less than the reinforcements at the tensile section.

iv) Then, the total cost of the beam (cost) per unit meter is calculated. It is the objective function of the design methodology, as given in equation [6.13]. If a constraint violation occurs, the cost is taken as 10^6 \$/m:

$$Cost = (A_g - A_s - A_s')C_c + (A_s + A_s')\gamma_s C_s \qquad [6.13]$$

The symbols denote the following definitions:

Cost = material cost of the beam per unit meter

A_g = gross area of the cross-section

A_s' = area of non-prestressed longitudinal compressive reinforcement

C_c = material cost of concrete per m^3

C_s = material cost of steel per ton

γ_s = specific gravity of steel

v) After the generation of the initial solutions, the modification of the candidate solutions is done according to the rules of the employed metaheuristic algorithm, and steps ii–iv are also considered. The process is done for a maximum iteration number.

The design variables are breadth (b_w), height (h), number of reinforcements in different sections (n_1, n_2, n_3 and n_4) and size of the reinforcement (ϕ_1, ϕ_2, ϕ_3 and ϕ_4), as shown in Figure 6.2.

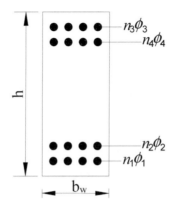

Figure 6.2. *RC beam with design variables*

The design constants of the optimization problem are the clear cover of the reinforcements, the maximum size of the aggregate diameter, the specified compressive strength of concrete, the specified yield strength of reinforcement, the diameter of the stirrup, and the cost of concrete and reinforcement bars, which were numerically taken as 35 mm, 16 mm, 20 MPa, 420 MPa, 10 mm, 40\$/m^3 and 400\$/ton, respectively. Discrete optimization was carried out because the constriction of concrete members cannot be done sensibly in a construction yard and the sizes of the reinforcements are fixed. Numerically, the design variable ranges were taken as 10–30 mm, 250–350 mm and 350–500 mm for the diameters of the main reinforcement bars, breadth and height, respectively. The dimensions were taken with 50 mm differences, and the reinforcement sizes are even numbers. All algorithms were done by taking 25 populations, and the switch probability of FPA was taken as 0.5.

The optimum results are presented in Tables 6.1–6.3 for different flexural moment goals. All algorithms are effective in finding the effective optimum values, but a perfect algorithm does not exist. The best performance is observed for the JA. It must be noted that all algorithms show different performances for different flexural moment goals. This situation validates the no-free-lunch theorem. According to the results, the best solutions can be found in the 20 runs, but minor local optima problems are observed for algorithms in several runs and the standard deviation values are not so small. The new generation algorithms are feasible, but the results and the robustness evaluation show that new algorithms or variants may be needed to improve existing progress.

Objective flexural moment (kNm)	50	100	150	200	250	300	350	400	450	500
h (mm)	350	400	500	500	500	500	500	500	500	500
b_w (mm)	250	250	250	250	250	300	300	350	350	400
ϕ_1 (mm)	14	16	16	28	26	22	26	26	28	26
ϕ_3 (mm)	20	28	30	22	14	12	16	12	16	12
n_1	2	3	4	2	3	5	4	5	5	6
n_3	0	0	0	0	2	2	3	6	5	9
ϕ_2 (mm)	12	10	12	10	12	14	12	10	10	12
ϕ_4 (mm)	18	16	14	16	18	22	18	26	14	30
n_2	2	4	2	3	2	2	4	3	2	4
n_4	0	0	0	0	0	0	0	0	0	0
M_u (kNm)	57.31	111.22	167.71	222.42	279.29	333.48	388.94	445.15	500.11	555.96
Cost ($/m) (best)	5.16	6.85	8.20	9.55	11.60	13.56	15.87	18.08	20.16	22.45
Number of analyses	225	100	75	100	125	200	1,475	950	175	2,100
Cost ($/m) (ave.)	5.17	6.86	8.25	9.59	11.65	13.70	16.05	18.62	20.58	22.93
Standard deviation	0.01	0.02	0.07	0.07	0.06	0.12	0.19	0.43	0.28	0.21

Table 6.1. *Optimum results of the RC beam (JA)*

Objective flexural moment (kNm)	50	100	150	200	250	300	350	400	450	500
h (mm)	350	400	500	500	500	500	500	500	500	500
b_w (mm)	250	250	250	250	250	250	300	300	350	400
ϕ_1 (mm)	14	16	16	28	26	28	26	28	28	26
ϕ_3 (mm)	30	30	16	20	14	14	14	16	16	12
n_1	2	3	4	2	3	3	4	4	5	6
n_3	0	0	0	0	2	4	4	5	5	9
ϕ_2 (mm)	12	10	12	10	12	12	14	12	12	12
ϕ_4 (mm)	28	14	16	10	18	18	18	12	16	20
n_2	2	4	2	3	2	3	3	4	2	4
n_4	0	0	0	0	0	0	0	0	0	0
M_u (kNm)	57.31	111.22	167.71	222.42	279.29	333.50	390.51	445.40	506.81	555.96
Cost ($/m) (best)	5.16	6.85	8.20	9.55	11.60	13.70	15.94	18.17	20.38	22.45
Number of analyses	925	50	475	625	1,125	100	1,125	925	350	900
Cost ($/m) (ave.)	5.16	6.85	8.23	9.56	11.71	13.87	16.15	18.43	20.68	22.97
Standard deviation	0.01	0.01	0.04	0.01	0.06	0.10	0.12	0.19	0.26	0.22

Table 6.2. Optimum results of the RC beam (TLBO)

Objective flexural moment (kNm)	50	100	150	200	250	300	350	400	450	500
h (mm)	350	400	500	500	500	500	500	500	500	500
b_w (mm)	250	250	250	250	250	300	300	450	400	450
ϕ_1 (mm)	14	16	16	28	26	22	26	20	24	26
ϕ_3 (mm)	24	18	28	16	14	10	20	22	14	14
n_1	2	3	4	2	3	5	4	8	6	6
n_3	0	0	0	0	2	3	2	0	5	5
ϕ_2 (mm)	12	10	12	10	12	14	12	12	10	10
ϕ_4 (mm)	20	18	26	18	14	30	24	28	22	24
n_2	2	4	2	3	2	2	4	4	7	6
n_4	0	0	0	0	0	0	0	0	0	0
M_u (kNm)	57.31	111.22	167.71	222.42	279.29	334.02	389.55	445.09	500.11	557.23
Cost ($/m) (best)	5.16	6.85	8.20	9.55	11.60	13.59	15.95	18.21	20.52	22.74
Number of analyses	675	1,225	1,950	1,450	450	1,575	1,400	1,150	2,025	2,225
Cost ($/m) (ave.)	5.30	6.87	8.25	9.56	11.78	13.95	16.30	18.60	20.84	23.07
Standard deviation	0.13	0.02	0.05	0.01	0.11	0.15	0.19	0.22	0.23	0.21

Table 6.3. Optimum results of the RC beam (FPA)

6.2. Optimum design of RC spread footings

Reinforced concrete (RC) spread footings are one of the most important components of structures since they direct structural loads to the ground. Thus, the design of spread footings is related to the interaction between soil and footings. In the optimum design, an economical and safe design under structural loads is not sufficient because of the stability against soil bearing capacity. For this reason, the problem is a complicated design that is highly nonlinear because of the existence of geotechnical and structural design constraints.

Different metaheuristic-based approaches have been developed for RC footings. Khajehzadeh *et al.* (2011) used a modified particle swarm optimization to find the optimum cost design of RC spread footings and retaining walls. Also, a gravitational search was proposed by Khajehzadeh *et al.* (2012) for shallow foundations. In addition to the cost objective, multi-objective optimizations considering CO_2 emissions have been developed by Camp and Assadallahi (2013) using a hybrid Big Bang–Big Crunch (BB-BC) algorithm, and Khajehzadeh *et al.* (2013) using a hybrid firefly algorithm.

As previously stated, the design of RC footings involves two different limit states called geotechnical and structural limit states. These states can be considered as different objectives. The objective function, which is the total cost of the RC footing, can be penalized with a big value if one of these limit states is not provided. If a violation occurs, the following analyses and design processes do not need to be conducted to minimize the computation time.

In this section, a design methodology for the optimum cost design of RC footings is presented and several design variables are randomized in different states of the methodology. Thus, it will be possible to stop the process or regenerate several design variables if several design variables are not physically applicable, and are not assigned. An example that can be given is the orientation of the bars. When the number of bars is assigned before, the assigned design variables may not be suitable to continue.

In the method presented, the detailed optimization includes design variables, such as the shape of the base of the footing, detailed reinforcements and height of the footing. The seven design variables are shown in Figure 6.3. X_1 and X_2 are the base dimensions, which are shown as L and B in the x and y directions, respectively. The third design variable (X_3) is the height of the footing (H). The sizes (diameters) of the reinforcement bars are shown as X_4 and X_5 in two directions, and the distances between the bars are shown as X_6 and X_7. The column may be under the effect of the flexural moment (M) and the axial force (P).

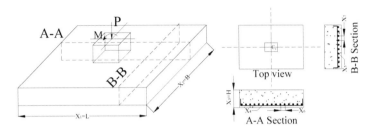

Figure 6.3. *Design variables of the optimization of RC spread footings*

The generation of design variables is done with a step by step procedure during the generation of the initial solutions and the modification of existing solutions. Firstly, the design variables about dimensions X_1–X_3 are defined. Then, the geotechnical limit states, such as the stress on the soil foundation and the settlement (δ), are checked.

The extremums of bearing pressure ($q_{max, min}$) at the sites of the foundation are found by using equation [6.14]. W_f is the total weight of the foundation including self-weight and the soil on the top of the foundation:

$$q_{min,max} = \frac{P + W_f}{BL} \pm \frac{6M}{BL^2} \qquad [6.14]$$

The minimum bearing pressure (q_{min}) must be compressive stress (over 0, as given in equation [6.15]) to ensure the stability of the foundation. In addition, the factor of safety (FS) is defined according to the maximum pressure (q_{max}), as given in equation [6.16]:

$$q_{min} \geq 0 \qquad [6.15]$$

$$FS < \frac{q_{ult}}{q_{max}} \qquad [6.16]$$

The ultimate bearing capacity of the soil (q_{ult}) is calculated in equation [6.17] for a cohesionless soil with no ground slope:

$$q_{ult} = \gamma D N_q F_{qs} F_{qd} + 0.5\gamma B N_\gamma F_{\gamma s} F_{\gamma qd} \qquad [6.17]$$

In equation [6.17], γ denotes the unit weight of the soil. The bearing capacity factors N_q and N_γ are shown in equations [6.18] and [6.19], respectively. The shape depth factors, F_{qs}, $F_{\gamma s}$, F_{qd}, $F_{\gamma d}$, are formulated in equations [6.20]–[6.23]. In these equations, ϕ' is the internal friction angle:

$$N_q = e^{\pi \tan \varphi'} \tan^2(\frac{\pi}{4} + \frac{\varphi'}{2}) \qquad [6.18]$$

$$N_\gamma = 2(N_q + 1)\tan \varphi' \qquad [6.19]$$

$$F_{qs} = 1 + \frac{B}{L}\tan \varphi' \qquad [6.20]$$

$$F_{\gamma s} = 1 - 0.4\frac{B}{L} \qquad [6.21]$$

$$F_{qd} = \begin{cases} 1 + 2\tan \varphi'(1 - \sin \varphi')^2 \left[\arctan(\frac{D}{B}) \right] & if\ D > B \\ 1 + 2\tan \varphi'(1 - \sin \varphi')^2 (\frac{D}{B}) & if\ D < B \end{cases} \qquad [6.22]$$

$$F_{\gamma d} = 1.0 \qquad [6.23]$$

The settlement can be calculated by using the elastic solution by Poulus and Davis (1974), as shown in equation [6.24]. The elasticity modulus and Poisson ratio of soil are denoted with E and v, respectively. The shape factor, denoted by β_z, is calculated using the equation by Whitman and Richart (1967), which is given in equation [6.25]:

$$\delta = \frac{P + W_f(1 - v^2)}{\beta_z E \sqrt{BL}} \qquad [6.24]$$

$$\beta_z = -0.0017(\frac{L}{B})^2 + 0.0597(\frac{L}{B}) + 0.9843 \qquad [6.25]$$

If one or more of the constraints that are given in equations [6.15], [6.16] and [6.24] are violated, the total cost, which is the objective function of the method, is equal to a big penalized value like 10^6 \$, and the process continues by getting through the generation or modification of the next solution. If not, the analyses will continue by assigning the reinforcements and checking the structural state limits.

The structural state limits involve checking the required flexural moment capacity in the critical sections of two directions, shear force capacity of the footing and two-way shear capacity (punching) of the footing. To check the structural state limits, the size and distance between the reinforcements are randomly defined and the three objectives considering structural state limits are checked using ACI-318 (2005). Similarly, the total cost is penalized with a big value when a possible violation occurs.

As they are only in control of structural state limits, the axial force and flexural moments are factored with different ϕ values for dead and live loads. ϕ is 1.4 and 1.7 for dead and live loads, respectively.

The critical flexural moments occur in the sections along the face of the column and the critical section is at the distance d_{ave} away from the face of the column. The average depth (d_{ave}) is the average of the effective depth of the foundation in two directions, since the effective depth is different in two directions.

The column tends to punch through the spread footing, and the check of the two-way shear force of the footing is necessary. The critical punching perimeter formulated in equation [6.26] is located $d_{ave}/2$ away from the column face. b_{column} and d_{column} denote the cross-sectional dimensions of the column

$$b_{perim} = 2b_{column} + 2d_{column} + 4d_{ave} \qquad [6.26]$$

In the flexural moment design, the analyses are similar to the design of the beam. The capacity of the RC footing for one-way shear ($V_{n,one-way}$) and two-way shear ($V_{n,two-way}$) are formulated in equations [6.27] and [6.28], respectively:

$$V_{n,one-way} = 0.75(0.17wd_{ave}\sqrt{f_c'}) \qquad [6.27]$$

$$V_{n,one-way} = 0.75 * \min \begin{cases} 0.17(1+\dfrac{2}{\beta})\sqrt{f'_c}\, b_{perim} d_{ave} \\[2mm] 0.083(\dfrac{4d_{ave}}{b_{perim}}+2)\sqrt{f'_c}\, b_{perim} d_{ave} \\[2mm] 0.33\sqrt{f'_c}\, b_{perim} d_{ave} \end{cases} \qquad [6.28]$$

The shear capacity is checked for two directions. The value of w is equal to B in the x direction and L for the y direction. The ratio of the long side to the short side of the column is shown by β. The compressive strength of concrete is denoted with f'_c. After checking all of the design constraints, the total cost of the footing is calculated in equation [6.29]

$$Cost = V_{concrete} C_c + W_{steel} C_s \qquad [6.29]$$

In equation [6.29], the objective function (cost) is calculated by the volume of concrete ($V_{concrete}$), the cost of concrete for the unit volume (C_c), the weight of the reinforcements (W_s) and the cost of reinforcement for the unit weight (C_s). The order and procedure are as defined above for the generation of initial solutions and the modification of existing solutions, by using algorithm rules.

As a numerical example, a spread footing supporting a column under axial force is optimized for cost minimization. The employed metaheuristic algorithms are FPA, TLBO and JA. The numerical values of design constants and ranges of design variables are given in Table 6.4.

The optimization was done for continuous and discrete design variables. The distances between the bars were randomized with a 10 mm difference. In the discrete optimization, the concrete dimensions are randomized with 0.05 m differences. The optimum results are presented for discrete and continuous design variables in Tables 6.5 and 6.6, respectively. The possible results of design variables are less in discrete optimization and TLBO and JA are effective at finding the same results with little computational effort. In continuous optimization, the best results for 20 runs are different for the algorithms.

Definition	Symbol	Unit	Value
Yield strength of steel	f_y	MPa	420
Compressive strength of concrete	f'_c	MPa	25
Concrete cover	c_c	mm	100
Max. aggregate diameter	D_{max}	mm	16
The elasticity modulus of steel	E_s	GPa	200
The specific gravity of steel	γ_s	t/m^3	7.86
The specific gravity of concrete	γ_c	kN/m^3	23.5
Cost of concrete per m^3	C_c	\$/m^3	40
Cost of steel per ton	C_s	\$/t	400
Internal friction angle of soil	ϕ'	°	35
Unit weight of base soil	γ_B	kN/m^3	18.5
Poisson ratio of soil	v	-	0.3
Modulus of elasticity of soil	E	MPa	50
Maximum allowable settlement	δ	mm	25
Factor of safety	FS	-	3.0
Minimum footing thickness	h_{min}	m	0.25
Column breadth in two directions	b/h	mm/mm	500/500
Dead axial loading	P_G	kN	750
Live axial loading	P_Q	kN	500
Range of width of the footing	B	m	2.0-5.0
Range of length of the footing	L	m	2.0-5.0
Range of height of the footing	H	m	0.25-1.0
Range of diameter of reinforcement bars	ϕ	mm	16-24
Range of distance between reinforcements	s	mm	5ϕ-250

Table 6.4. *Design constants and ranges of the design variables of RC spread footing*

	FPA	TLBO	JA
L (X_1) (m)	1.65	1.65	1.65
B (X_2) (m)	1.65	1.65	1.65
H (X_3) (m)	0.85	0.85	0.85
ϕ_x (X_4) (mm)	16	16	16
S_x (X_6) (mm)	250	250	250
ϕ_y (X_4) (mm)	16	16	16
S_y (X_7) (mm)	250	250	250
Best Cost ($)	104.92	104.92	104.92
Av. Cost ($)	105.80	104.92	104.92
Sta. Der. ($)	1.845	4.8×10^{-14}	7.2×10^{-14}
Analyses for optimum	2,961	84	67

Table 6.5. *Optimum results of RC spread footing with discrete variables*

	FPA	TLBO	JA
L (X_1) (m)	1.46	1.68	1.68
B (X_2) (m)	1.46	1.68	1.68
H (X_3) (m)	1.14	0.80	0.80
ϕ_x (X_4) (mm)	16	16	16
S_x (X_6) (mm)	250	250	250
ϕ_y (X_4) (mm)	16	16	16
S_y (X_7) (mm)	250	250	250
Best Cost ($)	107.73	103.02	103.22
Av. Cost ($)	108.03	103.04	105.00
Sta. Der. ($)	0.12	0.03	0.37
Analyses for optimum	1,267	2,783	1,284

Table 6.6. *Optimum results of RC spread footing with continuous variables*

6.3. Optimum design of RC columns

The vertical components of an RC structure, defined as columns, are under the combined effect of axial forces and flexural moments. Also, slenderness is an important factor for long columns, which must be designed according to second-order effects. In addition to the length of the columns, buckling parameters such as the cross-sectional area and the effective length factor in buckling (k) are also effective on second-order effects.

In design codes, the effect of slenderness or, in other words, second-order effects, are approximately considered with simple procedures, instead of the analysis of the additional second-order internal forces resulting from the deformed shape of the columns. Generally, the simple methods suggest the use of a factored moment according to the buckling behavior of the member.

In this section, the approximate design procedure given in ACI 318 (2005) is presented. The design moment is factored with a moment magnification factor (δ_s). In the calculation of the factor, the reduced moment of inertia of members is used to consider the cracking of members, and the reduction is 65% and 30% for beams and columns, respectively. Effective length factors in buckling (k) are found for all columns, and the Ψ value at both ends of the columns (Ψ_A at upper and Ψ_B at the lower end) is considered according to equation [6.30]. In this equation, E, I and l denote the elasticity modulus, the moment of inertia and the length of the RC members, respectively:

$$\Psi_{A,B} = \frac{\sum \left(\frac{EI}{l} \right)_{colum}}{\sum \left(\frac{EI}{l} \right)_{beam}} \qquad [6.30]$$

The value of k is obtained by using the following equations for a structural system, if the system is free to make lateral displacements

$$\Psi_A = 0.5(\Psi_A + \Psi_B) \qquad [6.31]$$

$$k = \frac{20 - \Psi_m}{20} \sqrt{1 + \Psi_m} \quad if\ \Psi_m < 2 \qquad [6.32]$$

$$k = 0.9\sqrt{1 + \Psi_m} \qquad if\ \Psi_m \geq 2 \qquad [6.33]$$

The moment magnification factor (δ_s) is formulated in equation [6.34]:

$$\delta_s = \frac{C_m}{1 - \dfrac{P_u}{0.75P_c}}$$

[6.34]

C_m denotes the correction factor considering moment diagram to end equivalent moment diagram and it is calculated by equation [6.35]. The value of C_m must be bigger than 0.4 and it is 1.0 if transverse loads exist between the joints

$$C_m = 0.6 + 0.4\frac{M_1}{M_2}$$

[6.35]

The critical buckling load can be calculated using to Euler's formula, by considering the reduction of the rigidity of the column by 75%. The critical buckling load formula by Euler is given in equation [6.36]:

$$P_c = \frac{\pi^2 EI}{(kl)^2}$$

[6.36]

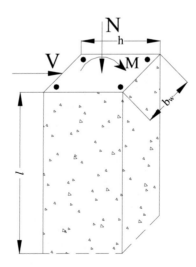

Figure 6.4. *Optimized RC column with loading conditions*

In the presented optimization procedure, the design constants are the length of the column (l), the clear cover (c_c), the maximum aggregate diameter (D_{max}), the elasticity modulus of steel (E_s), the specific gravity of steel (γ_s), the specific gravity of concrete (γ_c), the yield strength of steel (f_y), the compressive strength of concrete (f_c'), the cost of concrete per m³ (C_c) and the cost of steel per ton (C_s). Loading conditions such as axial force (N), shear force (V) and flexural moment (M) are also defined. The loading of the column is shown in Figure 6.4.

The design variables of the optimization problem are the breadth of column (b_w), the height of the column (h), the number and diameter size of the longitudinal reinforcement bars in two lines (including web reinforcements) and the diameter size and distance of shear reinforcements, as shown in Figure 6.5. The design of the longitudinal reinforcements is symmetrical. For this reason, the assignments of candidate solutions for a section represent both the upper and lower sections of the column.

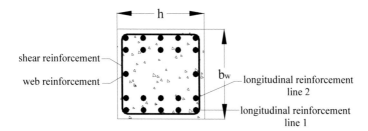

Figure 6.5. *Design variables of the optimum RC column problem*

The generation of design variables starts with cross-sectional dimensions. Since columns are under compressive axial forces, the yielding of steel before the cracking of concrete may not be provided. In this case, the conditions are given in equation [6.37], and equation [6.38] must be provided to ensure the ductile behavior of the member. In equations [6.37] and [6.38], A_c denotes the cross-sectional area ($b_w h$) of the column. In the methodology, b_w and h are iteratively randomized until these conditions are provided

$$V < \begin{cases} 0.2 f_c' A_c \\ 5.5 A_c \end{cases} \qquad [6.37]$$

$$N < 0.5 f_c' A_c \qquad\qquad\qquad [6.38]$$

Then, the generation of reinforcement sizes and numbers is carried out. The orientation of reinforcement bars is checked according to the ACI-318 rules given in equation [6.39]. According to those rules, the methodology can position the reinforcements in two lines. The clear distance between the reinforcements (a_ϕ) is related to the average of the diameter sizes in a line ($\phi_{average}$) and the maximum aggregate diameter (D_{max}). a_ϕ is the clear distance between the reinforcements. The reinforcements are iteratively randomized until the placement condition is satisfied

$$a_\phi > \begin{cases} 1.5\phi_{avarage} \\ 40\ mm \\ \dfrac{4}{3} D_{max} \end{cases} \qquad\qquad [6.39]$$

The methodology includes the generation of web reinforcements. Also, the minimum and maximum reinforcement conditions are checked. In this case, the reinforcement ratio (ρ), calculated by considering all longitudinal reinforcements, must be between 0.01 and 0.06. The iterative generations will continue until the limit states about the reinforcements are provided.

After the design constraints are provided, the distance from the extreme compression fiber to the neural axis (c) is scanned for the required axial force capacity. Thus, the flexural moment capacity of the candidate solution is found. The moment capacity must exceed the moment goal and it must also not be too high from the goal for an economical solution. For this reason, the iterative generations of candidate solutions are done if the flexural moment capacity is lower than the required one, or more than a defined percentage of the required value. In the numerical example presented, this percentage is taken as 100% and is iteratively increased by 1% for every 500 iterations to prevent the trapping of the methodology.

Lastly, the design of shear reinforcements was carried out. Since the shear design is not complicated compared to the combined effect of axial force and bending moment, the diameter sizes are iteratively assigned with values within range and the required distance of shear reinforcement (stirrups) is found by checking all possible values for minimum cost. The nominal shear strength of concrete (V_c) and the nominal shear strength of reinforcement (V_s) are given in equations [6.40] and [6.41], respectively:

$$V_c = \frac{\sqrt{f_c'}}{6} b_w d$$ [6.40]

$$V_s = \frac{A_v f_y d}{s}$$ [6.41]

A_v and s denote the shear reinforcement area and the distance between them. The effective depth of concrete is shown by d. In addition, the V_s value must not exceed $0.66\sqrt{f_c'} b_w d$. Also, the minimum shear reinforcement ($A_{v,min}$) value and maximum shear reinforcement distance (s_{max}), respectively, defined in equations [6.42] and [6.43] are checked. The objective function is penalized with a very big value if the required shear safety is not provided

$$(A_v)_{min} = \frac{1}{3} \frac{b_w s}{f_y}$$ [6.42]

$$s_{max} \begin{cases} \leq \dfrac{d}{4} & if\ V_s \geq 0.33\sqrt{f_c'} b_w d \\ \leq \dfrac{d}{2} & if\ not \end{cases}$$ [6.43]

After a candidate solution is found, the objective function of the optimization is calculated. It is the maximum material cost (C) given in equation [6.44], which is minimized. The parameters of the objective function are presented in Table 6.7:

$$\min C = (A_g - A_{st})C_c + (A_{st} + \frac{A_v}{s} u_{st})l\gamma_s C$$ [6.44]

Definition	Symbol
Material cost of the beam per unit meter	C
Gross area of cross-section	A_g
Area of non-prestressed longitudinal reinforcement	A_{st}
Area of shear reinforcement spacing s	A_v
Length of shear reinforcement spacing s	u_{st}
Material cost of concrete per m^3	C_c
Material cost of steel per ton	C_s
Specific gravity of steel	γ_s

Table 6.7. *Parameters of the objective function of the optimum RC column*

The above methodology is used for the generation of a candidate solution. The algorithm rules apply to the generation and modification of solution matrices. The above methodology has been used with several metaheuristic algorithms, such as the HS (Bekdaş and Nigdeli 2012), TLBO (Bekdaş and Nigdeli 2016a) and Bat algorithm (BA) (Bekdaş and Nigdeli 2016b). The results of these algorithms are presented as a numerical example. The numerical values of the design constants are shown in Table 6.8.

Definition	Value
Range of web width, b_w	250–400 mm
Range of height, h	300–600 mm
Longitudinal reinforcement (ϕ)	16–30 mm
Shear reinforcement (ϕ_v)	8–14 mm
Effective length factor in buckling, k	1.2
Clear cover, c_c	30 mm
Max. aggregate diameter, D_{max}	16 mm
Yield strength of steel, f_y	420 MPa
Comp. strength of concrete, f'_c	25 MPa
Elasticity modulus of steel, E_s	200,000 MPa
Specific gravity of steel, γ_s	7.86 t/m^3
Specific gravity of concrete (γ_c)	2.5 t/m^3
Cost of concrete per m^3	40 $
Cost of steel per ton	400$

Table 6.8. *Design constants and ranges of design variables*

The RC column is subjected to 2,000 kN, 50 kNm and 50 kN internal forces for axial force, flexural moment and shear force, respectively. To show the effect of slenderness, the optimization problem is presented for different lengths of the column and the results are presented in Table 6.9. As observed from the results, the increase in the total cost does not show a linear increase for long columns, because of the slenderness and magnified flexural moment. With the increase in the cross-sectional dimensions, a reduction in the number of shear reinforcements is also observed. TLBO is effective in finding the same result as BA, and TLBO has no user-defined parameters different from the population of the class. For simplification, TLBO is the most feasible for the problem.

	HS			BA			TLBO		
Length of the column (l)	3 m	4 m	5 m	3 m	4 m	5 m	3 m	4 m	5 m
Breadth of the column (b_w) (mm)	400	300	300	400	300	300	400	300	300
Height of the column (h) (mm)	400	550	600	400	550	600	400	550	600
Bars in each face	1Φ20+1Φ18	2Φ16	2Φ16	3Φ16	2Φ16	2Φ16	3Φ16	2Φ16	2Φ16
Web reinforcement in each face	1Φ18	1Φ16+1Φ18	2Φ18	1Φ16	1Φ16+1Φ18	2Φ18	1Φ16	1Φ16+1Φ18	2Φ18
Shear reinforcement diameter (mm)	Φ8	Φ8	Φ8	Φ8	Φ8	Φ8	Φ8	Φ8	Φ8
Shear reinforcement distance (mm)	170	240	270	170	240	270	170	240	270
Optimum cost ($)	38.58	52.27	69.97	38.22	52.27	69.97	38.22	52.27	69.97

Table 6.9. *Optimum values of the RC column problem*

6.4. Optimum design of RC frames

The frame structures are formed from beam and column elements. The optimization of RC frames contains the optimization of both the beam and the frames. Also, the element rigidities related to the dimensions of elements are effective in the solution of internal force and responses of the structure for statically indetermined structures. In this case, the optimization methodology must involve the analysis of the structure.

Several metaheuristic-based methods have been proposed for RC frames. The GA-based method by Rajeev and Krishnamoorthy (1998) includes a detailed reinforcement design. Camp *et al.* (2003) considered the slenderness of columns in a GA-based RC frame optimization approach and presented a two-bay six-story frame example by grouping cross-sections of beams and columns. Lee and Ahn (2003) used a database for the possible design of members in a GA-based RC frame optimization methodology, by considering the static lateral equivalent earthquake loads. Govindaraj and Ramasamy (2007) also proposed a GA using a methodology for statically loaded RC frames. Another classical algorithm, SA, was used by Paya *et al.* (2008) for the multi-objective optimization of RC frames. Also, minimum embedded CO_2 emissions have been considered with cost by using SA (Paya-Zaforteza *et al.* 2009) and BB-BC (Camp and Huq 2013) for RC frames. The music-inspired HS has also been used for RC frames for static (Akin and Saka 2012, 2015) and dynamic (Nigdeli and Bekdaş 2016a; Bekdaş and Nigdeli 2017a) analyses. Kaveh and Sabzi (2011) combined several metaheuristic algorithms for the optimum design of RC frames.

In this section, the improved HS methodology for RC frames (Nigdeli and Bekdaş 2016b; Bekdaş and Nigdeli 2017b) is presented. The presented methodology improved with several random stages because of the optimization of various types of structural members with different design constraints and materials. In classical approaches, the optimum cross-sectional dimensions are combined with random reinforcement designs, which are far from an optimum solution. When several limitations and groupings of design variables ensuring design constraints are carried out, partial effectiveness can be observed for classical metaheuristic methods, but a general optimization cannot be provided. For this reason, the presented methodology uses sub-optimization stages with an additional iterative random search.

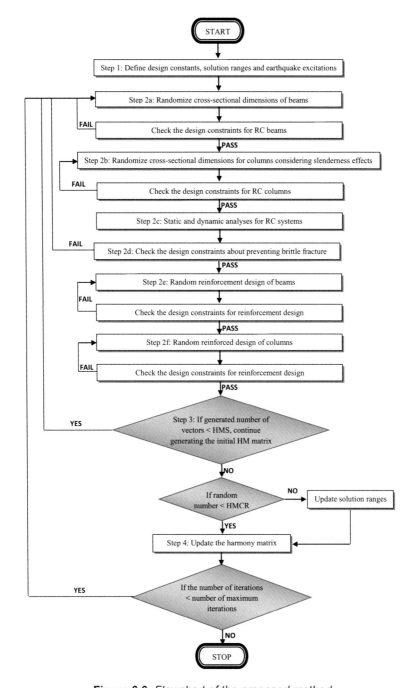

Figure 6.6. *Flowchart of the proposed method*

The presented methodology considers the time history analyses of earthquake records instead of using equivalent static lateral forces. The design constraints and rules from ACI-318 (2005) were previously presented for beams and columns. The methodology is summarized in the flowchart given in Figure 6.6. The random search stages are combined with HS, which is described in Chapter 3. The static and dynamic analyses of the structure are carried out after the dimensions of the elements are generated. After the critical internal forces are found, the check of the other constraints and the generation of variables about the reinforcements is carried out. The methodology can also be used with another algorithm. The presented methodology considers the unfavorable loading of Live (L) loads. The distributed loads may be chosen as equally distributed, triangular distributed or trapezium distributed, as shown in Figure 6.7.

Figure 6.7. *Types of load distributions for the RC frame*

The depth (d) and breadth (b_w) of beams must provide the following statements, according to ACI 318, if l is the length and h is the height of the beam

$$d < \frac{1}{4} \qquad\qquad\qquad [6.45]$$

$$b_w \geq 0.3h . \qquad\qquad\qquad [6.46]$$

The following statement must also be provided for the beams if b is the breadth of the supporting column

$$b_w \leq b + \frac{3}{2}d \qquad\qquad\qquad [6.47]$$

Different from the design constants of the RC beam and column examples, the nodal points of elements are also design constants. In the methodology, static and dynamic responses are calculated by using the stiffness method and time history analyses. In these analyses, the mass and stiffness matrices of the system are constructed by merging the element matrices in global coordinates. The frame structure is idealized as a shear building. The mass (M_e) and stiffness (K_e) matrices of a structural member are shown in equations [6.48] and [6.49], respectively. In these element matrices, D, L, g, E, l, I, A and n are the dead loads, live loads, gravity, elasticity modulus, length, moment of inertia, cross-sectional area of the element and live load participation factor, respectively:

$$M_e = \frac{\gamma_c Al + (D + nL)(l - a)}{420g}\begin{bmatrix} 140 & 0 & 0 & 70 & 0 & 0 \\ 0 & 156 & 22l & 0 & 54 & -13l \\ 0 & 22l & 4l^2 & 0 & 13l & -3l^2 \\ 70 & 0 & 0 & 140 & 0 & 0 \\ 0 & 54 & 13l & 0 & 156 & -22l \\ 0 & -13l & -3l^2 & 0 & -22l & 4l^2 \end{bmatrix} \quad [6.48]$$

$$K_e = \begin{bmatrix} \frac{EA}{l} & 0 & 0 & -\frac{EA}{l} & 0 & 0 \\ 0 & \frac{12EI}{l^3} & \frac{6EI}{l^2} & 0 & -\frac{12EI}{l^3} & \frac{6EI}{l^2} \\ 0 & \frac{6EI}{l^2} & \frac{4EI}{l} & 0 & -\frac{6EI}{l^2} & \frac{2EI}{l} \\ -\frac{EA}{l} & 0 & 0 & \frac{EA}{l} & 0 & 0 \\ 0 & -\frac{12EI}{l^3} & -\frac{6EI}{l^2} & 0 & \frac{12EI}{l^3} & -\frac{6EI}{l^2} \\ 0 & \frac{6EI}{l^2} & \frac{2EI}{l} & 0 & -\frac{6EI}{l^2} & \frac{4EI}{l} \end{bmatrix} \quad [6.49]$$

In the dynamic analyses, the equation of motion subjected to ground acceleration is written as

$$M\ddot{x}(t) + C\dot{x}(t) + Kx(t) = -M\{1\}\ddot{x}_g(t) \quad [6.50]$$

where x(t) is the deflection vector including displacements and rotations, C is the damping matrix calculated by assuming 5% inherent damping for all modes for RC structures, [1] is a vector of ones with elements as much as the degrees of freedom of the structure and $\ddot{x}_g(t)$ is the ground acceleration. The first and second derivatives of x(t) (velocity and acceleration) with respect to time are shown as $\dot{x}(t)$ and $\ddot{x}(t)$, respectively. Several earthquake records are used for the analyses, and the number of earthquakes can be modified according to the design codes and region of the structure. The internal forces are calculated by multiplying the stiffness matrix with the deflections and the dynamic responses are divided into the elastic response parameter (R). The design internal forces (U) are calculated for the most critical earthquake data and time according to equations [6.51]–[6.53]. The unfavorable loading of live actions and every time lag of earthquake force (E) are considered in the analyses and the most critical load is taken into account:

$$U = 1.4D + 1.7L \tag{6.51}$$

$$U = 0.75(1.4D + 1.7L) \pm E \tag{6.52}$$

$$U = 0.9D \pm E \tag{6.53}$$

Similarly, the total cost of the RC frame is taken as the objective function (OF), which is the sum of the cost of the elements for a system (C_e) with n elements, as given in equation [6.54]. When a constraint violation occurs, it is penalized with a big value like the beam and column examples

$$OF = \sum_{i=1}^{n} \left(C_e \right)_i \tag{6.54}$$

The methodology is presented with two frame structures, which are a two-span two-story symmetrical RC frame and a three-span three-story RC frame. The information about the three different earthquake records used is given in Table 6.10. The numerical values of the design constants and the ranges are shown in Table 6.11, and discrete variables are used for the practical application of the design. The design variables about the dimensions are assigned to the values which are the multiples of 50 mm, and the diameters of reinforcement bars are assigned to the values which are the multiples of 2 mm. For the distributed loads, the ratio of a and l is taken as ¼. The same constant values were taken for both frame structures and only the coordinates of the nodal points are different.

Earthquake	Date	Station	Component
Imperial Valley	1940	117 El Centro	I-ELC180
Northridge	1994	24514 Sylmar	SYL360
Loma Prieta	1989	16 LGPC	LGP000

Table 6.10. *Earthquake records used in the RC frame*

Note: Earthquake records were taken from the PEER NGA DATABASE (http://peer.berkeley.edu/nga/) (2005)

Definition	Symbol	Unit	Value
Range of web width	b_w	mm	250–400
Range of height	h	mm	300–600
Clear cover	c_c	mm	30
Range of reinforcement	ϕ	mm	16–30
Range of shear reinforcement	ϕ_v	mm	8–14
Max. aggregate diameter	D_{max}	mm	16
Yield strength of steel	f_y	MPa	420
Comp. strength of concrete	f'_c	MPa	25
Elasticity modulus of steel	E_s	MPa	200,000
Specific gravity of steel	γ_s	t/m^3	7.86
Specific gravity of concrete	γ_c	t/m^3	2.5
Elastic response parameter	R	-	8.5
Cost of concrete per m^3	C_c	$	40
Cost of steel per ton	C_s	$	400

Table 6.11. *Design constants and ranges of design variables of RC frame examples*

6.4.1. *The first example: two-span two-story RC frame*

The model of a two-span two-story RC frame is shown in Figure 6.8. The optimum results of the first frame example are shown in Table 6.12 for the RC column members, and in Table 6.13 for the RC beam members. For the beams, LJ and RJ represent the left and right joints of the element. The total optimum cost of the first RC frame is 304.78 $.

Element number	b_w (mm)	h (mm)	Bars in each face	Shear reinforcement diameter/distance (mm)
1	250	300	2Φ10+ 2Φ12	Φ8/120
2	250	300	2Φ10+ 2Φ12	Φ8/120
3	250	300	2Φ10+ 2Φ12	Φ8/120
6	250	300	2Φ10+ 2Φ12	Φ8/120
7	250	300	2Φ10+ 2Φ12	Φ8/120
8	250	300	2Φ10+ 2Φ12	Φ8/120

Table 6.12. *Optimum design of columns (Frame 1)*

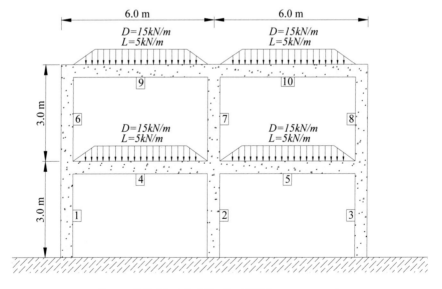

Figure 6.8. *Model of the first RC frame example*

According to the optimum results, the columns of the first numerical example are the same because the dynamic earthquake forces are the critical loading condition of the structure. Although the system is symmetrical and the analyses can be shortened by solving the half system, the optimization is done by taking the full system to see the accuracy of the methodology. The effectiveness of the methodology has been proved since the optimum results of the symmetric columns and beams are the same.

Element number	b_w (mm)	h (mm)	Bars in comp. section	Bars in tensile section	Shear reinforcement diameter/distance (mm)
LJ4			1Φ18+1Φ14 +1Φ16	1Φ20+1Φ30+1Φ14	
4	250	300	2Φ12	1Φ20+1Φ16	Φ8/120
RJ4–LJ5			2Φ16+1Φ12 +1Φ14	1Φ24+1Φ18+1Φ28	
5	250	300	2Φ12	1Φ20+1Φ16	Φ8/120
RJ5			1Φ18+1Φ14 +1Φ16	1Φ20+1Φ30+1Φ14	
LJ9			1Φ18+ 1Φ16	1Φ12+2Φ14+1Φ24	
9	250	300	2Φ12	1Φ14+1Φ26	Φ8/120
RJ9–LJ10			1Φ16+1Φ14 +1Φ20	1Φ22+1Φ24+1Φ16+1Φ18	
10	250	300	2Φ12	1Φ14+1Φ26	Φ8/120
LJ10			1Φ18+ 1Φ16	1Φ12+2Φ14+1Φ24	

Table 6.13. *Optimum design of beams (Frame 1)*

6.4.2. *The first example: two-span two-story RC frame*

The three-span three-story structure is the second numerical example, and the model is shown in Figure 6.9. The optimum results of the second frame are presented in Tables 6.14 and 6.15 for column and beam elements, respectively. The total cost of the optimum design is 724.54 $, and it is nearly half of an engineered design (Bekdaş and Nigdeli 2017a).

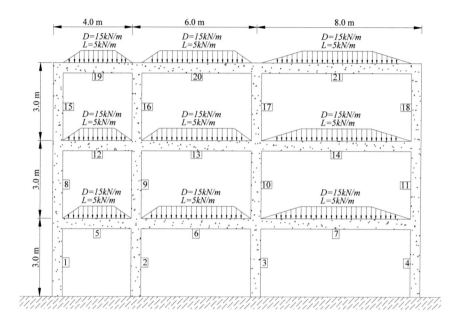

Figure 6.9. *Model of the second RC frame example*

Element number	b_w (mm)	h (mm)	Bars in each face	Shear reinforcement diameter/distance (mm)
1	250	300	2Φ10+ 2Φ12	Φ8/120
2	250	350	3Φ10+ 1Φ12	Φ8/150
3	250	350	1Φ12+ 1Φ18+1Φ10	Φ8/150
4	250	300	2Φ10+ 2Φ12	Φ8/120
8	250	300	2Φ10+ 2Φ12	Φ8/120
9	250	300	2Φ10+ 2Φ12	Φ8/120
10	250	300	4Φ10+ 1Φ12	Φ8/120
11	250	300	2Φ10+ 2Φ12	Φ8/120
15	250	300	2Φ10+ 2Φ12	Φ8/120
16	250	300	2Φ10+ 2Φ12	Φ8/120
17	250	300	1Φ14+ 2Φ12	Φ8/120
18	250	300	3Φ18	Φ8/120

Table 6.14. *Optimum design of columns (Frame 2)*

Element number	b_w (mm)	h (mm)	Bars in comp. section	Bars in tensile section	Shear reinforcement diameter/distance (mm)
LJ5			1Φ10+2Φ14 +1Φ16	1Φ24+1Φ28	
5	250	350	2Φ10+ 1Φ12	1Φ14+1Φ16+1Φ12	Φ8/150
RJ5–LJ6			1Φ10+1Φ26	1Φ16+ 2Φ14+1Φ28	
6	250	350	2Φ10+ 1Φ12	1Φ20+ 1Φ14	Φ8/150
RJ6–LJ7			1Φ22+1Φ14 +1Φ18	1Φ20+1Φ18+1Φ26+1Φ22	
7	250	450	1Φ10+1Φ12 +1Φ14	1Φ24+ 1Φ20	Φ8/200
RJ7			1Φ20+ 1Φ22	1Φ14+1Φ16+1Φ28+1Φ20	
LJ12			1Φ20+ 1Φ12	1Φ30+ 1Φ12	
12	250	450	1Φ16+ 1Φ14	1Φ12+ 1Φ22	Φ8/200
RJ12–LJ13			4Φ12	1Φ28+ 1Φ18	
13	250	400	4Φ10	1Φ14+ 1Φ12+1Φ16	Φ8/170
RJ13–LJ14			1Φ28+ 1Φ14	2Φ28+ 1Φ18	
14	250	400	1Φ16+ 1Φ12	1Φ10+ 2Φ12+1Φ26	Φ8/170
RJ14			1Φ16+1Φ22 +1Φ12	1Φ10+1Φ22+1Φ28+1Φ16	
LJ19			1Φ16+ 1Φ12	1Φ10+ 1Φ18+1Φ14	
19	250	400	2Φ14	2Φ14	Φ8/170
RJ19–LJ20			1Φ12+ 1Φ22	1Φ10+ 2Φ22+1Φ12	
20	250	300	2Φ12	1Φ10+1Φ16+1Φ12+1Φ14	Φ8/120
RJ20–LJ21			1Φ24+1Φ18	1Φ18+1Φ26+1Φ24+1Φ12	
21	250	400	1Φ12+ 1Φ16	1Φ24+ 1Φ26	Φ8/170
LJ21			1Φ20+ 1Φ12	1Φ26+ 1Φ20	

Table 6.15. *Optimum design of beams (Frame 2)*

6.5. Optimum design of RC cylindrical walls

The liquid tanks are formed by cylindrical walls. There are different metaheuristic-based methodologies for RC cylindrical walls for various design objectives. The optimum design, including the dimension and reinforcements, can be carried out, or post-tension forces may be optimized to reduce the internal forces. The general option is to optimize post-tension forces, dimension and reinforcement together.

6.5.1. Optimum design of axially symmetric RC walls

The design of RC cylindrical walls has been investigated by Bekdaş (2014) by using the HS algorithm and optimizing the design variables, such as thickness (h), the compressive strength of concrete (f'_c) and the size/distance of reinforcements in vertical and horizontal directions. The objective of the method is to minimize the total cost of the wall ($C(x)$), which is given in equation [6.55]:

$$\min C(x) = C_c V_c(x) + C_s W_s(x) + C_{fw} A_{fw}(x) \qquad [6.55]$$

In equation [6.55], C_c, C_s and C_{fw} denote the unit costs for concrete, steel and formwork, respectively. The concrete volume of the wall, the total weight of the reinforcements, and the surface area of the wall are shown with V_c, W_s and A_{fw}, respectively.

The model of the cylindrical wall used as a water tank is shown in Figure 6.10. The radius and the height of the tank are shown by using r and H, respectively.

The design constraints of the optimization problem are summarized in Table 6.16. In this table, S_v and S_h represent the distance between the centroid of the bars in vertical and horizontal directions, respectively. M_u, V_u and T_u represent the ultimate moment, shear forces and axial tension. A_s^{hoop}, A_{sv} and A_{sh} are the areas of hoop reinforcing bars, horizontal bars and vertical bars, respectively. The yield strength of steel is denoted by f_y. ϕ is the strength reduction factor described in ACI-318 (2005).

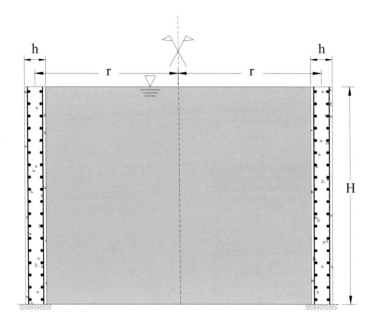

Figure 6.10. *Model of the cylindrical wall*

Description	Constraints
Flexural strength capacity, M_d	$g_1(X)$: $M_d \geq M_u$
Shear strength capacity, V_d	$g_2(X)$: $V_d \geq V_u$
The axial tension, T	$g_3(X)$: $T = \phi A_s^{hoop} f_y \geq T_u$
Minimum steel area, A_{smin}	$g_4(X)$: $A_{sv} \geq A_{smin}$ and $A_{sh} \geq A_{smin}$
Maximum crack width, w_{max}	$g_5(X)$: $w_{max} \leq 0.1$ mm
Maximum steel bar spacing, S_{max}	$g_6(X)$: $S_v \leq S_{max}$ and $S_h \leq S_{max}$
Minimum steel bar spacing, S_{min}	$g_7(X)$: $S_v \geq S_{min}$ and $S_h \geq S_{min}$
Minimum concrete cover, c_{cmin}	$g_8(X)$: $c_{cmin} \geq 40$ mm

Table 6.16. *Constraints of the RC cylindrical wall*

The general flow of the methodology is shown in Figure 6.11. In this methodology, the analyses of the cylindrical wall are done using the superposition method (SPM) developed by Hetenyi (1946).

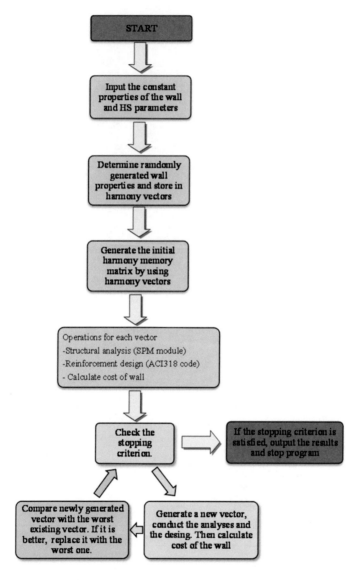

Figure 6.11. *Flowchart for the optimum design of the RC cylindrical wall. For a color version of this figure, see www.iste.co.uk/toklu/metaheuristics.zip*

6.5.2. *Optimization of post-tensioning forces for cylindrical walls*

Post-tensioned cables can be used on cylindrical walls to reduce the flexural moment of the wall. Thus, cracking will be prevented, and the design will be more secure compared to the wall without post-tensioning forces. Bekdaş and Nigdeli (2017b) used several metaheuristics for the problem. The new generation algorithms such as the HS, Bat algorithm (BA), flower pollination algorithm (FPA), and teaching–learning-based optimization (TLBO) have been used in the methodology combined with SPM (Hetenyi 1946). The model of the cylindrical wall with post-tensioning forces is shown in Figure 6.12.

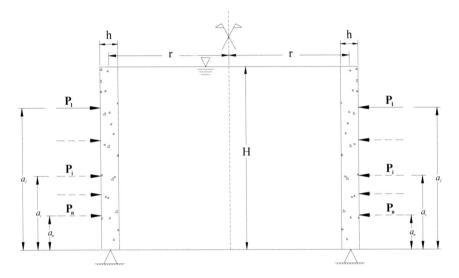

Figure 6.12. *The model of the post-tensioned cylindrical wall*

The design variables are the intensities (P_1 to P_n) and the distance (a_1 to a_n) of the forces. The objective function is minimizing the maximum flexural moment ($M_{w\text{-}PT}(x)$), as given in equation [6.56]

$$\min M_{W-PT}(x) = M_{W/O-PT}(x) + \sum_{i=1}^{n} M_{LMP_i}(x) \qquad [6.56]$$

It is obtained by SPM, and the values of the solutions with and without post-tensioning forces are added. Since the moments are in a different direction, there is a reduction in the total moment. The $M_{w/o\text{-}PT}(x)$ and $M_{LMPi}(x)$

represent the moment for the wall without post-tensioning forces and the moment that occurred because of ith post-tensioning forces, respectively.

In this section, an optimization problem for the optimization of 10 post-tensioning forces is presented. The design constants of the numerical example are 18.25 m, 0.3 m, 4 m, 2,060,100 kN/m^2 and 0.15 for the radius (r), the thickness of the wall (h), height (H), elasticity modulus (E) and Poisson's ratio (v), respectively. The wall is simply supported from the base and is filled with water. In this case, the density of the liquid is 9.81 kN/m^3. The maximum intensity limit for the force is 78.48 kN/m and this value is two times the water load on the wall. The optimum results of the problem are presented in Table 6.17 for different algorithms. The change in the longitudinal moment along the wall height can be seen for different algorithms in Figure 6.13.

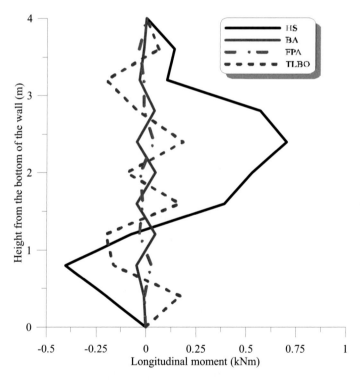

Figure 6.13. *Longitudinal moment along the wall height. For a color version of this figure, see www.iste.co.uk/toklu/metaheuristics.zip*

	HS		BA		FPA		TLBO	
	P_i (kN)	a_i (m)	P_i (kN)	a_i (m)	P_i (kN)	a_i (m)	P_i (kN)	a_i (m)
	6.1800	0.0800	9.8895	0.5136	5.0976	0.0000	54.0380	0.0351
	0.6400	0.1700	29.9384	0.6492	37.5497	0.0043	15.7862	0.0616
	12.1000	0.3600	39.2094	1.3612	17.9924	0.0069	7.6956	0.5083
	10.8100	0.7500	40.0533	1.8878	16.2979	0.3310	22.8662	1.0064
	15.1600	1.0900	63.6585	2.5964	20.9789	0.9752	10.8170	1.7591
	10.8700	1.7000	10.7418	3.0882	2.3922	1.1193	5.5239	1.9272
	7.7900	2.1700	55.4566	3.3912	17.1457	1.7842	7.1614	2.5476
	3.3500	2.6500	13.3034	3.5743	0.0000	1.7972	0.0315	2.8346
	0.2000	3.0700	3.0223	3.9799	10.4187	2.5770	0.1237	2.9771
	4.0200	3.1200	18.8184	3.9818	4.2293	3.3336	5.0336	3.0760
Best Moment (kNm/m)	0.7000		0.0454		0.0369		0.2032	
Average Moment (kNm/m)	1.27445		0.04616		0.11512		0.46382	
Standard deviation	0.18450		0.00081		0.06059		0.19302	
Number of analyses	696,107		167,614		213,617		39,818	

Table 6.17. *Optimum results for post-tensioned RC wall*

The FPA is the most effective in the reduction of the moment of the numerical problem. The most robust one is BA, while HS and TLBO are not effective on the optimum reduction of the maximum moment along with the wall height.

6.5.3. *Optimization of post-tensioned axially symmetric cylindrical RC walls*

Bekdaş proposed the HS-based optimization procedure for optimization of the design of the RC wall and post-tensioning force together. The objective of the optimization ($f(x)$) is also the minimization of the total cost of the wall, including post-tensioning, as formulated in equation [6.57]. The objective function is dependent on the design variables presented in Table 6.18. The unit cost of concrete, steel, post-tensioning and formwork are shown by C_c, C_s, C_{pt} and C_{fw}, respectively

$$\min f(X) = C_c(x_{2n+2})V_c(x_{2n+1}) + C_sW_s(x_{2n+3}, x_{2n+4}, x_{2n+5}, x_{2n+6}) + $$
$$+ C_{pt}W_{pt}(x_1, \ldots x_{2n}) + C_{fw}A_{fw}$$

[6.57]

Description	Design variable
Post-tensioning loads	$x_1=P_1, x_2=P_2, ... x_n=P_n$
Locations of post-tensioning loads	$x_{n+1}=a_1, x_{n+2}=a_2, ... x_{2n}=a_n$
Thickness of the wall, h	$x_{2n+1}=h$
Compressive strength of concrete, f'_c	$x_{2n+2}=f'_c$
Diameter of vertical reinforcing bars, ϕ_v	$x_{2n+3}=\phi_v$
Diameter of horizontal reinforcing bars, ϕ_h	$x_{2n+4}=\phi_h$
Distance between vertical reinforcing bars, S_v	$x_{2n+5}=S_v$
Distance between horizontal reinforcing bars, S_h	$x_{2n+6}=S_h$

Table 6.18. *Design variables of the optimization of a post-tensioned wall*

$V_c\ (x_{2n+1})$, $W_s(x_{2n+3}, x_{2n+4}, x_{2n+5}, x_{2n+6})$, $W_{pt}(x_1,...x_{2n})$ and A_{fw} represent the wall volume, the total weight of reinforcements, the total weight of post-tensioning cables and the surface area of the wall, respectively. The problem has the same design constraints as the example presented in section 6.5.1. Also, the analyses are done using SPM (Hetenyi 1946) in the optimization methodology.

Bekdaş (2015) concluded that post-tensioned cylindrical RC walls are more economical than walls without post-tensioning, if the height of the wall exceeds a limit. For short walls, post-tensioning is not necessary to save on the cost of the tank.

7

Optimization in Structural Control Problems

In civil engineering, structures subjected to unsteady excitations can be controlled by specific systems, including passive, active, semi-active and hybrid control systems. Generally, these are used in towers, bridges, off-shore structures and high-rise buildings. In structural control applications, the control systems are used to absorb or minimize the vibrations of different sources.

One of the major applications of structural control systems is in earthquake motions. During earthquakes, the bearing elements of structures may be damaged and this may lead to the collapse or loss of the structure for future use. Additionally, the sensing of a strong motion may be unconfident for the people or non-structural members. This may lead to damage being caused or the loss of precious material like historical valuables or valuable equipment. Also, serviceability of the structure may be prevented. For this reason, hospitals and nuclear power plants are the most important buildings to use control systems.

In this chapter, three different structural control applications using metaheuristic-based optimization techniques are presented. The applications include optimum parameter tuning of tuned mass dampers (TMDs) and optimization of base isolation systems. Two different TMD optimization methodologies are presented and metaheuristic methods are adopted with time and frequency history analyses in methodologies.

7.1. Optimum design of tuned mass dampers (TMD)

TMD was invented to absorb undesired vibrations by using tuned properties of its mechanical components. The optimum tuning of TMDs relies on the structural properties which are period and damping of the structure. The factor of excitation may play a significant role in the performance and the numerical values of the optimum TMD parameters. The usage area of TMDs is dynamic systems such as vibrating machines, all types of vehicles including automotive, trains and ships, robotic devices and all types of civil structures exposed to dynamic excitations such as earthquakes, strong winds, vehicle or pedestrian traffic and water. In that case, the examples of practical applications are found in the constructed high-rise buildings, towers, vibration-sensitive structures like nuclear plants, offshore structures and motorway bridges, viaducts and footbridges.

The most well-known application is the sphere-shaped TMD of the Taipei 101 building in Taiwan, installed during construction to reduce both wind and earthquake induced vibrations. Generally, TMDs have been installed in structures after their construction to improve strength and comfort. For example, a 1.5 t TMD suspended by three cables and four hydraulic telescopic shock absorbers was installed on the top of the Berlin TV Tower in Germany (Figure 7.1) to reduce wind induced vibrations. A seismic retrofit example is in California, USA. The theme building positioned in Los Angeles International Airport (LAX) (Figure 7.2) was modified by using a TMD positioned on the top of the concrete core as a slab-like structure supported by eight fluid viscous dampers and isolation systems. A response reduction between 30% and 40% is achieved in the numerical analyses for the TMD, which has a 20% mass of the main structure (Miyamoto *et al.* 2011).

The basic working principle of TMDs for single degree of freedom (SDOF) structures is to add a secondary mode to the structural system and the TMD is tuned with a frequency that is too close to the natural frequency of the main structural system. Thus, the system combined with TMD will have two frequencies which are a little lower and higher than an SDOF frequency For this reason, the resonance case seen when the excitation frequency and the SDOF natural frequency are matched is prevented. Additionally, the excitation is more effective on the system when the excitation frequency is too close to the natural frequency of the system and the essential purpose in structures is that. The damping of TMDs is effective in the reduction of vibrations resulting from excitations with various frequencies that cover wind and ground excitations resulting from earthquakes.

Figure 7.1. *The Berlin TV Tower. For a color version of this figure, see www.iste.co.uk/toklu/metaheuristics.zip*

Figure 7.2. *The Theme Building at LAX (Miyamoto et al. 2011). For a color version of this figure, see www.iste.co.uk/toklu/metaheuristics.zip*

The first ancestor of TMDs is the device invented by Frahm (1911) which contains a mass attached to a spring-like stiffness element. Ormondroyd and Den Hartog (1947) added additional damping devices to absorb vibrations resulting from random frequency excitations.

The estimation of TMD parameters has been investigated and many methods have been developed. It is still an active research area since an exact formulation cannot be derived for damped structures exposed to excitations with random frequencies. The exciting formulations are obtained according to several assumptions and numerical trials. The first and well-known optimum tuning equations of TMD are the formulations presented by Den Hartog (1947). The equations of Den Hartog are for finding the optimum frequency ratio (f_{opt}) (the ratio of frequencies of TMD, $\omega_{d,opt}$, and main structure, ω_s) and the optimum damping ratio ($\xi_{d,opt}$) which is the ratio of the damping coefficient of TMD (c_d) and twice the mass of the damper (m_d) multiplied by $\omega_{d,opt}$, respectively. The equations of Den Hartog are proposed for undamped SDOF main systems. The inherent damping of the main system is also considered in several studies (Bishop and Welbourn 1952; Snowdon 1959; Falcon et al. 1967; Ioi and Ikeda 1978). Warburton (1982) also suggested equations for different excitations like white noise. The numerical algorithms and investigations were used to develop equations for systems with inherent damping (ξ) by Sadek et al. (1997) and a curve fitting method was used for several numerical solutions. Also, the application of these equations can be approximately done for multiple degrees of freedom (MDOF) systems by considering the critical mode (Warburton and Ayorinde 1980). Several formations concerning a mass ratio of TMD and SDOF structures (μ) are presented in Table 7.1.

The only way to investigate all of the vibration modes is to use numerical algorithms and metaheuristic-based methods which are the most important proposal for TMDs in the last two decades. The employed metaheuristic algorithms include the genetic algorithm (GA) (Hadi and Arfiadi 1998; Singh et al. 2002; Desu et al. 2006; Pourzeynali et al. 2007; Marano et al. 2010), particle swarm optimization (PSO) (Leung et al. 2008) including close form expressions presented in Table 7.1 (Leung and Zhang 2008), the bionic algorithm (Steinbuch 2011), harmony search (HS) (Bekdaş and Nigdeli 2017a; Bekdaş et al. 2017b; Nigdeli and Bekdaş 2017c; Nigdeli et al. 2017; Zhang and Zhang 2017), ant colony optimization (ACO) (Farshidianfar and Soheili 2013a), the artificial bee colony algorithm (ABC) (Farshidianfar and Soheili 2013b), the flower pollination algorithm (FPA) (Bekdaş et al. 2017a; Nigdeli et al. 2017a) and teaching–learning-based optimization (TLBO) (Nigdeli and Bekdaş 2015; Bekdaş et al. 2017b; Nigdeli et al. 2017b).

Method	$f_{opt} = \dfrac{w_{d,opt}}{w_s}$	$\xi_{d,opt} = \dfrac{c_{d,opt}}{2m_d w_{d,opt}}$
Den Hartog (1947)	$\dfrac{1}{1+\mu}$	$\sqrt{\dfrac{3\mu}{8(1+\mu)}}$
Warburton (1982)	$\dfrac{\sqrt{1-(\mu/2)}}{1+\mu}$	$\sqrt{\dfrac{\mu(1-\mu/4)}{4(1+\mu)(1-\mu/2)}}$
Sadek et al. (1997)	$\dfrac{1}{1+\mu}\left[1-\xi\sqrt{\dfrac{\mu}{1+\mu}}\right]$	$\dfrac{\xi}{1+\mu}+\sqrt{\dfrac{\mu}{1+\mu}}$
Leung & Zhang (2009)	$\dfrac{\sqrt{1-(\mu/2)}}{1+\mu}$ $+(-4.9453+20.2319\sqrt{\mu}-37.9419\mu)\sqrt{\mu}\xi$ $+(-4.8287+25.0000\sqrt{\mu})\sqrt{\mu}\xi^2$	$\sqrt{\dfrac{\mu(1-\mu/4)}{4(1+\mu)(1-\mu/2)}}$ $-5.3024\xi^2\mu$

Table 7.1. *Frequency and damping ratio expressions of the TMD optimization*

In this chapter, two types of methodology will be presented for metaheuristic methods. In the first, the results of the time domain analyses are considered in the main objective of the optimization, while the frequency domain analyses are used in the objective for the second.

7.1.1. *Time domain-based optimization of TMDs*

The presented TMD optimization methodology considers the passive control of SDOF structures for ground excitation. A physical model of an SDOF main structure with a TMD is shown in Figure 7.3. The structure slab is taken as a rigid mass (m) in the shear building representation. The stiffness of the structure is shown with k and it is found for the force needed for the unit displacement of the structure. The inherent damping of structures results from the internal friction of cracks and is represented with a dashpot element with a damping coefficient, c. The freedom of the SDOF structure is x, while x_d denotes the displacement of TMD with respect to the ground. The TMD is represented with a mass (m_d), a spring, and dashpot which have stiffness and a damping coefficient of TMD equal to k_d and c_d, respectively.

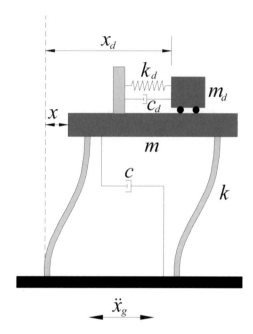

Figure 7.3. *The physical model of the SDOF structure with TMD*

The equation of motion of the SDOF structure without a TMD is formulated in equation [7.1]. The ground excitation is shown as $\ddot{x}_g$. A dot on x represents a derivative of x with respect to time. If double dots are positioned on x, the second derivative of x, with respect to time, is shown. The SDOF structure with TMD has two coupled equations, given in equations [7.2] and [7.3]. The solutions of dynamic analyses are found with a solving method for all iterations:

$$m\ddot{x} + c\dot{x} + kx = -m\ddot{x}_g \tag{7.1}$$

$$m\ddot{x} + (c + c_d)\dot{x} - c_d\dot{x}_d + (k + k_d)x - k_dx_d = -m\ddot{x}_g \tag{7.2}$$

$$m_d\ddot{x}_d - c_d\dot{x} + c_d\dot{x}_d - k_d\dot{x} + k_d\dot{x}_d = -m\ddot{x}_g \tag{7.3}$$

The period (T_d) and damping ratio (ξ_d) of TMD are calculated according to the following equations, and these properties and μ are the design variables of the presented methodology and numerical example:

$$T_d = 2\pi \sqrt{\frac{m_d}{k_d}} \qquad\qquad [7.4]$$

$$\xi_d = \frac{c_d}{2\sqrt{m_d k_d}} \qquad\qquad [7.5]$$

After the design constants (structural properties, design variable limits, the data of excitation and specific algorithm parameters), candidate solutions are randomly chosen to generate an initial solution matrix. For all sets of design variables of the initial matrix, the dynamic analyses are conducted and the objective function values are stored. The objective function (OF) is the maximum solution of x in absolute value as given in equation [7.6]:

$$OF=\min (\max|x|) \qquad\qquad [7.6]$$

The optimum design also has two design constraints. The value of the OF is penalized by adding the violation value to the analysis solution. The iterations finish when the maximum number of iterations is reached.

The first design constraint (g_1) is a limitation of the stroke capacity of TMD. As given in equation [7.7], the normalized maximum stroke by using the displacement of the structure without TMD must be lower than the maximum stroke capacity (stmax). The stroke capacity has an effect on optimum parameters and a high damping value is required to decrease the displacement of the TMD.

The second constraint (g_2) is controlled according to the frequency (ω) response of the structure. The maximum amplitude of the transfer function ($TF(\omega)$) of a story acceleration must be less than or equal to the value of the structure without TMD:

$$g_1 = \frac{\max(|x_d - x|)_{withTMD}}{\max(x)_{withoutTMD}} \leq st - max \qquad\qquad [7.7]$$

$$g_2 = \max |(TF(w))|_{withTMD} \leq \max |(TF(w))|_{withotTMD} \qquad\qquad [7.8]$$

In the optimization process, the generated initial solution matrix is modified according to the newly-generated candidate solution by using formulations of the metaheuristic algorithms for all iterations. By eliminating the worst one from the existing and modified solutions, the optimum TMD parameters are found.

In the presented numerical investigation, six different pulse-like ground motions (three different periods of two types) are used in the optimization proceeds and the most critical excitation with the maximum value is considered. The pulse-like ground motions are seen in near-fault ground motions and these motions are more dangerous to the structure because of these significant impulsive pulses with a big peak ground velocity (V_p) and long-period (T_p) (or small frequency; ω_p). These pulses are namely the flint step that is parallel to the fault rupture and the directivity pulse that is perpendicular to the fault rupture. These pulses are presented in Figure 7.4 for the ground displacement (Steward et al. 2001). These pulses were modeled with the trigonometric functions of Makris (1997). The formulations of type A and type B motions are given in equations [7.9] and [7.10]; these formulations are with respect to time (t) and types A and B represent the flint step and directivity pulse, respectively,

$$\ddot{x}_g(t)=\omega_p \frac{V_p}{2}\sin(\omega_p t) \qquad 0 \leq t \leq T_p \qquad\qquad [7.9]$$

$$x_g(t)=\frac{V_p}{2}t-\frac{V_p}{2\omega_p}\sin(\omega_p t) \qquad 0 \leq t \leq T_p \qquad\qquad [7.10]$$

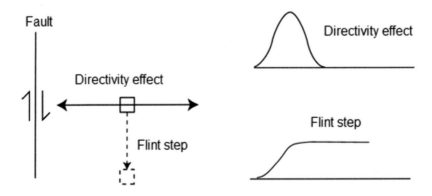

Figure 7.4. *Impulsive motions*

An SDOF system with a 2 s vibration period (T) and a 5% damping ratio (ξ) was investigated. The periods of impulsive motions were taken as 1.5 s, 2 s and 2.5 s which are around and equal to the period of the structure. The V_p was taken as 200 m/s. The TMD was limited to a stmax value taken as 2. The mass of the structure (m) was assumed as 1 kg to find an optimum TMD, which has the same mass and mass ratio value. The stiffness and damping coefficients of the structure are calculated according to equations [7.11] and [7.12]:

$$k = (\frac{2\pi}{T})^2 m = \frac{4\pi^2 m}{T^2} \qquad\qquad [7.11]$$

$$c = 2\xi \sqrt{km} \qquad\qquad [7.12]$$

The optimum results found by using four different metaheuristic algorithms are presented. These algorithms are harmony search (HS), flower pollination algorithm (FPA), teaching–learning-based optimization (TLBO) and Jaya algorithm (JA). The population number and harmony memory considering rate were taken as 5. The maximum number of analyses is 1,000, and the algorithms are tested for the repeating optimization process 30 times. All other algorithm-specific parameters were taken as 0.5. The range of μ was defined between 0.01 and 0.2 to keep the axial force level of columns in the desired range. For easy optimization, the range of T_d was defined between 0.5 and 1.5 times the period of the superstructure. The damping ratio of TMD (ξ_d) was searched between 0.01 and 0.3. The final optimum results are given in Table 7.2.

	JA	FPA	TLBO	HS
μ	0.2000	0.2000	0.1999	0.1987
T_d	1.9286	1.9287	1.9303	1.9801
ξ_d	0.0473	0.0473	0.0479	0.0681
OF_{best}	1.2089	1.2089	1.2091	1.2149
$OF_{average}$	1.2241	1.2407	1.2118	1.2249
Standard deviation	0.0305	0.0340	0.0032	0.0053

Table 7.2. *Optimum results*

The maximum displacement under the critical excitation, which is the directivity pulse with a 2 s period, is 1.602 m and this value is reduced to 1.20888 m, 1.208901 m, 1.20911 m and 1.21489 m for JA, FPA, TLBO and HS, respectively. In this case, all algorithms are accepted as effective on the optimization, but the JA algorithm outperforms the other algorithms in the minimization of OF.

7.1.2. Frequency domain-based optimization of TMDs

In the frequency-based optimization of TMDs, MDOF main structures are considered and the formulations are presented. A TMD positioned on the shear building model is shown in Figure 7.5. The equation of motion of the shear building with a TMD is formulated in equation [7.13] in matrix form:

$$\mathbf{M}\ddot{x}(t) + \mathbf{C}\dot{x}(t) + \mathbf{K}x(t) = -\mathbf{M}\{1\}\ddot{x}_g(t)$$

[7.13]

The mass (M), damping (C) and stiffness (K) matrices of a shear building with a TMD on the top are given in equations [7.14][7.16]. The vector of the response of the structure $(y(t))$, given in equation [7.17], includes the displacement values of all stories $(x_i(t)$ for $i=1, 2, 3, ..., N$ if N is the number of stories) and TMD $(x_d(t))$ with respect to the ground, and structural responses occur due to the ground acceleration $(\ddot{x}_g(t))$. The dot (.) placed on $y(t)$ represents the derivation of responses with respect to time and $\ddot{x}(t)$ and $\dot{x}(t)$ are the representations of acceleration and velocity of the structure with respect to the ground, respectively.

$$M = \text{diag}[m_1\, m_2\, ...\, m_N\, m_d]$$

[7.14]

$$C = \begin{bmatrix} (c_1 + c_2) & -c_2 & & & & & \\ -c_2 & (c_2 + c_3) & -c_3 & & & & \\ & & \cdot & \cdot & & & \\ & & \cdot & \cdot & \cdot & & \\ & & & \cdot & \cdot & \cdot & \\ & & & & -c_N & (c_N + c_d) & -c_d \\ & & & & & -c_d & c_d \end{bmatrix}$$

[7.15]

$$K = \begin{bmatrix} (k_1 + k_2) & -k_2 & & & & & \\ -k_2 & (k_2 + k_3) & -k_3 & & & & \\ & & \cdot & & \cdot & & \\ & & & \cdot & & \cdot & \\ & & & & \cdot & \cdot & \\ & & & & -k_N & (k_N + k_d) & -k_d \\ & & & & & -k_d & k_d \end{bmatrix} \qquad [7.16]$$

$$x(t) = \begin{Bmatrix} x_1(t) \\ x_2(t) \\ \vdots \\ x_N(t) \\ x_d(t) \end{Bmatrix} \qquad [7.17]$$

The notifications such as m, c and k represent mass, damping and stiffness coefficients, respectively, and the subscripts of these values represent the story number. The properties of the TMD are shown as m_d, c_d and k_d by using the subscript, d.

The transfer function is defined as the ratio of Laplace transformations of a structural response (acceleration in the current methodology) to the external excitation (the ground acceleration in the current methodology). The acceleration transfer function (TF(w)), which is a function of frequency (w), is formulated in equation [7.18]. The TF(ω) vector contains the values of all stories (TF$_1$, TF$_2$, ..., TF$_N$) and TMD (TF$_d$):

$$TF(w) = \begin{bmatrix} TF_1(w) \\ TF_2(w) \\ \vdots \\ TF_N(w) \\ TF_d(w) \end{bmatrix} = \left[-Mw^2 + Cwj + K \right]^{-1} Mw^2 \mathbf{1} \qquad [7.18]$$

The objective function $f_i(x_i)$ of the frequency-based method is to minimize the amplitude of the acceleration transfer function of the top story by finding the best set of design variables (x_i). The function $f_i(x_i)$ is formulated in equation [7.19] for the ith candidate solution:

$$f_i(x_i) = 20 \, Log_{10} \left| \max\left(TF_N(w)\right)\right| \qquad [7.19]$$

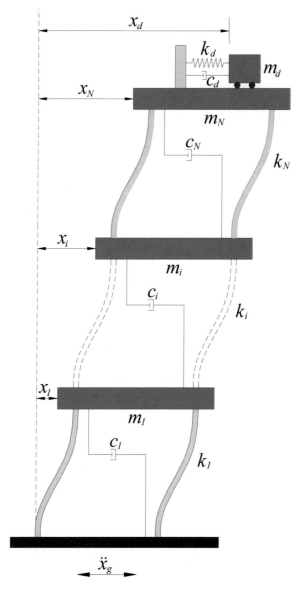

Figure 7.5. *MDOF shear building model with TMD*

The transfer function contains real and imaginary parts and the absolute value is the function taken to consider the amplitude value. The unitless transfer function is transformed to decibels (dB) by taking the base-10 logarithm of the amplitude and multiplying by 20. The design variables of the methodology are mass (m_d), period (T_d) and damping ratio (ξ_d) of TMD, as given in equation [7.20]:

$$x_i = \begin{Bmatrix} m_{di} \\ T_{di} \\ \xi_{di} \end{Bmatrix} \quad for\ i = 1, 2, ..., n \qquad [7.20]$$

The optimization steps are the same as the time domain-based method. The only differences are the objective function and the type of analysis. In this method, the objective function can be directly found by using the formula given in equation [7.18], and the use of ground acceleration data is not needed.

The numerical examples involve two shear buildings which are 10-story and 40-story structural models. The employed metaheuristic algorithms are JA, HS, FPA and TLBO. The chosen population number for the algorithms (HMS for HS) is 25. The other specific parameters such as HMCR, PAR and switch probability are taken as 0.5, 0.2 and 0.5, respectively.

The ranges of the design variables are as follows:

– The mass of the TMD (m_d) must be between 1% and 10% of the total mass of the structures. The total mass of the 40-story structure may be too much, for that reason, the maximum mass ratio is taken as 2% for a second case.

– The period of the TMD (T_d) is searched between 0.8 and 1.2 times of the critical period of the structure.

The damping ratio of TMD (ξ_d) must be between 1% and 30%.

7.1.2.1. Optimum results for a 10-story structure

The properties of the 10-story shear building model (Singh 1997) are presented in Table 7.3.

Story	m_i (t)	k_i (MN/m)	c_i (MNs/m)
1	360	650	6.2
2	360	650	6.2
3	360	650	6.2
4	360	650	6.2
5	360	650	6.2
6	360	650	6.2
7	360	650	6.2
8	360	650	6.2
9	360	650	6.2
10	360	650	6.2

Table 7.3. *Properties of the 10-story structure*

The critical natural frequency of the 10-story structure is 1 Hz, and the value of maximum acceleration transfer function amplitude is 26.2091 dB for the structure without TMD. By using the optimum TMD parameters presented in Table 7.4, the maximum amplitude of the acceleration transfer function is reduced to 11.3316 dB. The results of all algorithms are also evaluated for 20 runs of the algorithms, and the average values (f_{ave}), the standard deviation (σ) and the number of variable evaluations needed to reach the optimum results (c_{best}) are calculated. It can be seen that the JA and the FPA outperform the other algorithms by means of computational effort.

	HS	FPA	TLBO	JA
m_d (t)	360	360	360	360
T_d (s)	1.109193	1.108639	1.108639	1.108639
ξ_d	0.262109	0.25969	0.25969	0.25969
f_{best}	11.33419	11.33157	11.33157	11.33157
f_{ave}	11.33295	11.33157	11.33157	11.33157
σ	0.000321	0	3.62E-13	0
c_{best}	7,006	950	5,750	1,575

Table 7.4. *Optimum results of the 10-story structure*

The acceleration transfer function of the top story of the 10-story structure is plotted in Figure 7.6. In addition to the reduction in the first peak value, the optimum TMD also has a minor effect on the second peak in frequency of the second mode.

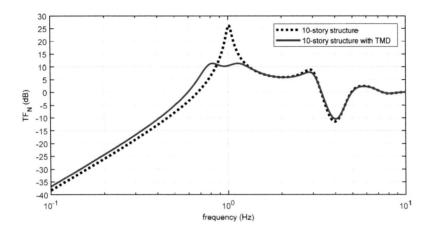

Figure 7.6. *TF$_N$ plot for the 10-story structure*

7.1.2.2. Optimum results for a 40-story structure

The proposed methodology was also tested for a big structure with 40 stories. The properties taken from Liu *et al.* (2008) are given in Table 7.5. As mentioned, the optimum TMD parameters were found for two cases of maximum mass ratio. In cases 1 and 2, the maximum allowable mass ratio of the TMD is 10% and 2%, respectively. The optimum values for the 40-story structure are presented in Table 7.6. JA is the fastest algorithm in finding the optimum results of the TMD parameters of the 40-story structure.

In case 1, the optimum mass of TMD is bigger than the mass of a story of the structure. In this case, case 2 will be more realistic than case 1 in practical application. The maximum amplitude seen for 0.25 Hz (4 s period) is 26.68 dB for the structure without TMD. The maximum amplitude is reduced to 11.7273 dB and 17.3285 dB for case 1 and case 2, respectively. The frequency plot for the top-story acceleration transfer function is seen in Figure 7.7 for algorithms FPA, TLBO and JA.

Story	m_i (t)	k_i (MN/m)	c_i (MNs/m)
1	980	2130.00	42.60
2	980	2100.97	42.02
3	980	2071.95	41.44
4	980	2042.92	40.86
5	980	2013.90	40.28
6	980	1984.87	39.70
7	980	1955.85	39.12
8	980	1926.82	38.54
9	980	1897.79	37.96
10	980	1868.77	37.38
11	980	1839.74	36.79
12	980	1810.72	36.21
13	980	1781.69	35.63
14	980	1752.67	35.05
15	980	1723.64	34.47
16	980	1694.62	33.89
17	980	1665.59	33.31
18	980	1636.56	32.73
19	980	1607.54	32.15
20	980	1578.51	31.57
21	980	1549.49	30.99
22	980	1520.46	30.41
23	980	1491.44	29.83
24	980	1462.41	29.25
25	980	1433.38	28.67
26	980	1404.36	28.09
27	980	1375.33	27.51
28	980	1346.31	26.93
29	980	1317.28	26.35
30	980	1288.26	25.77

Story	m_i (t)	k_i (MN/m)	c_i (MNs/m)
31	980	1259.23	25.18
32	980	1230.21	24.60
33	980	1201.18	24.02
34	980	1172.15	23.44
35	980	1143.13	22.86
36	980	1114.10	22.28
37	980	1085.08	21.70
38	980	1056.05	21.12
39	980	1027.03	20.54
40	980	998.00	19.96

Table 7.5. *Story parameters of the 40-story structure*

		HS	FPA	TLBO	JA
Case 1	m_d (t)	3920	3,920	3,920	3,920
	T_d (s)	4.35918	4.358596	4.358596	4.358596
	ξ_d	0.281695	0.281294	0.281294	0.281294
	f_{best}	11.72978	11.72735	11.72735	11.72735
	f_{ave}	11.72975	11.72866	11.72735	11.72735
	σ	0.000472	0.002619	0	0
	c_{best}	44,349	1,725	5,150	1,375
Case 2	m_d (t)	784	784	784	784
	T_d (s)	3.94892	3.949575	3.949575	3.949575
	ξ_d	0.131255	0.131671	0.131671	0.131671
	f_{best}	17.33442	17.32852	17.32852	17.32852
	f_{ave}	17.33572	17.32852	17.32852	17.32852
	σ	0.002938	3.65E-15	3.65E-15	3.65E-15
	c_{best}	43,186	1,450	4,800	1,300

Table 7.6. *Optimum results of the 40-story structure*

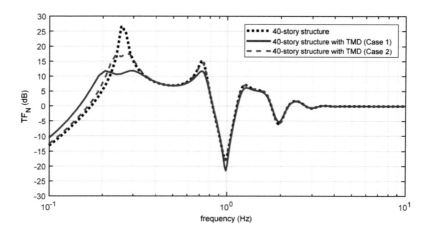

Figure 7.7. *TF$_N$ plot for the 40-story structure. For a color version of this figure, see www.iste.co.uk/toklu/metaheuristics.zip*

Case 1 is also effective on the maximum responses of the second and third natural frequencies of the structure, as seen in the plot. Case 2 is only effective at the first natural frequency, but it is an economical and feasible application.

7.1.2.3. Optimum results for a 40-story structure

It is seen that the optimum TMDs are effective in the reduction of the acceleration transfer function at the points where peak amplitude is seen. The main purpose of TMD is to damp vibrations by reducing the maximum displacement and providing a rapid steady-state response. For that reason, the optimum results obtained by JA were tested using several groups of earthquake data. These data are found in FEMA P-695: Quantification of Building Seismic Performance Factors (2009). The earthquake records in Table 7.7 represent the far-field ground motions.

The optimum TMD parameters were also tested under the near-field ground motions including (Table 7.8) or not including (Table 7.9) significant impulsive pulses. Tables 7.7–7.9 contain the record information, including earthquake name, recording station, year, magnitude, fault normal (FN) and fault parallel (FP) components.

Earthquake no.	Earthquake name	Recording station	Year	Magnitude	FN component	FP component
1	Northridge	Beverly Hills - Mulhol	1994	6.7	NORTHR/MUL009	NORTHR/MUL279
2	Northridge	Canyon Country-WLC	1994	6.7	NORTHR/LOS000	NORTHR/LOS270
3	Duzce, Turkey	Bolu	1999	7.1	DUZCE/BOL000	DUZCE/BOL090
4	Hector Mine	Hector	1999	7.1	HECTOR/HEC000	HECTOR/HEC090
5	Imperial Valley	Delta	1979	6.5	IMPVALL/H-DLT262	IMPVALL/H-DLT352
6	Imperial Valley	El Centro Array #11	1979	6.5	IMPVALL/H-E11140	IMPVALL/H-E11230
7	Kobe, Japan	Nishi-Akashi	1995	6.9	KOBE/NIS000	KOBE/NIS090
8	Kobe, Japan	Shin-Osaka	1995	6.9	KOBE/SHI000	KOBE/SHI090
9	Kocaeli, Turkey	Duzce	1999	7.5	KOCAELI/DZC180	KOCAELI/DZC270
10	Kocaeli, Turkey	Arcelik	1999	7.5	KOCAELI/ARC000	KOCAELI/ARC090
11	Landers	Yermo Fire Station	1992	7.3	LANDERS/YER270	LANDERS/YER360
12	Landers	Coolwater	1992	7.3	LANDERS/CLW-LN	LANDERS/CLW-TR
13	Loma Prieta	Capitola	1989	6.9	LOMAP/CAP000	LOMAP/CAP090
14	Loma Prieta	Gilroy Array #3	1989	6.9	LOMAP/G03000	LOMAP/G03090
15	Manjil, Iran	Abbar	1990	7.4	MANJIL/ABBAR—L	MANJIL/ABBAR—T
16	Superstition Hills	El Centro Imp. Co.	1987	6.5	SUPERST/B-ICC000	SUPERST/B-ICC090
17	Superstition Hills	Poe Road (temp)	1987	6.5	SUPERST/B-POE270	SUPERST/B-POE360
18	Cape Mendocino	Rio Dell Overpass	1992	7.0	CAPEMEND/RIO270	CAPEMEND/RIO360
19	Chi-Chi, Taiwan	CHY101	1999	7.6	CHICHI/CHY101-E	CHICHI/CHY101-N
20	Chi-Chi, Taiwan	TCU045	1999	7.6	CHICHI/TCU045-E	CHICHI/TCU045-N
21	San Fernando	LA Hollywood Stor	1971	6.6	SFERN/PEL090	SFERN/PEL180

Earthquake no.	Earthquake name	Recording station	Year	Magnitude	FN component	FP component
22	Friuli, Italy	Tolmezzo	1976	6.5	FRIULI/A-TMZ000	FRIULI/A-TMZ270

Table 7.7. *Far-field ground motions (FEMA P-695 2009)*

Earthquake no.	Earthquake name	Recording station	Year	Magnitude	FN component	FP component
1	Imperial Valley-06	El Centro Array #6	1979	6.5	Impvall/H-E06_233	Impvall/H-E06_323
2	Imperial Valley-06	El Centro Array #7	1979	6.5	Impvall/H-E07_233	Impvall/H-E07_323
3	Irpinia, Italy-01	Sturno	1980	6.9	Italy/A-Stu_223	Italy/A-Stu_313
4	Superstition Hills-02	Parachute Test Site	1987	6.5	Superst/B-Pts_037	Superst/B-Pts_127
5	Loma Prieta	Saratoga - Aloha	1989	6.9	Lomap/Stg_038	Lomap/Stg_128
6	Erzican, Turkey	Erzican	1992	6.7	Erzikan/Erz_032	Erzikan/Erz_122
7	Cape Mendocino	Petrolia	1992	7.0	Capemend/Pet_260	Capemend/Pet_350
8	Landers	Lucerne	1992	7.3	Landers/Lcn_239	Landers/Lcn_329
9	Northridge-01	01 Rinaldi Receiving Sta	1994	6.7	Northr/Rrs_032	Northr/Rrs_122
10	Northridge-01	01 Sylmar - Olive View	1994	6.7	Northr/Syl_032	Northr/Syl_122
11	Kocaeli, Turkey	Izmit	1999	7.5	Kocaeli/Izt_180	Kocaeli/Izt_270
12	Chi-Chi, Taiwan	TCU065	1999	7.6	Chichi/Tcu065_272	Chichi/Tcu065_002
13	Chi-Chi, Taiwan	TCU102	1999	7.6	Chichi/Tcu102_278	Chichi/Tcu102_008
14	Duzce, Turkey	Duzce	1999	7.1	Duzce/Dzc_172	Duzce/Dzc_262

Table 7.8. *Near-field ground motions with pulses (FEMA P-695 2009)*

Earthquake no.	Earthquake name	Recording station	Year	Magnitude	FN component	FP component
1	Gazli, Ussr	Karakyr	1976	6.8	Gazli/Gaz_177	Gazli/Gaz_267
2	Imperial Valley-06	El Centro Array #7	1979	6.5	Impvall/H-Bcr_233	Impvall/H-Bcr_323
3	Imperial Valley-06	Sturno	1979	6.5	Impvall/H-Chi_233	Impvall/H-Chi_323
4	Nahanni, Canada	Site 1	1985	6.8	Nahanni/S1_070	Nahanni/S1_160
5	Nahanni, Canada	Site 2	1985	6.8	Nahanni/S2_070	Nahanni/S2_160
6	Loma Prieta	Bran	1989	6.9	Lomap/Brn_038	Lomap/Brn_128
7	Loma Prieta	Corralitos	1989	6.9	Lomap/Cls_038	Lomap/Cls_128
8	Cape Mendocino	Cape Mendocino	1992	7.0	Capemend/Cpm_260	Capemend/Cpm_350
9	Northridge-01	La - Sepulveda Va	1994	6.7	Northr/0637_032	Northr/0637_122
10	Northridge-01	Northridge - Saticoy	1994	6.7	Northr/Stc_032	Northr/Stc_122
11	Kocaeli, Turkey	Yarimca	1999	7.5	Kocaeli/Ypt_180	Kocaeli/Ypt_270
12	Chi-Chi, Taiwan	Tcu067	1999	7.6	Chichi/Tcu067_285	Chichi/Tcu067_015
13	Chi-Chi, Taiwan	Tcu084	1999	7.6	Chichi/Tcu084_271	Chichi/Tcu084_001
14	Denali, Alaska	Taps Pump Sta. #10	2002	7.9	Denali/Ps10_199	Denali/Ps10_289

Table 7.9. *Near-field ground motions without pulses (FEMA P-695 2009)*

The structural responses obtained from the 10-story structure with and without TMD are presented in Tables 7.10–7.12 for far-field records, the near-field records with and without pulses, respectively. The maximum values given in Tables 7.10–7.12 are the top-story displacement (x_{10}) and the total acceleration of the top story ($\ddot{x}_{10} + \ddot{x}_g$). The FP component (DUZCE/BOL090) of the Bolu record of the 1999 Düzce earthquake is the most critical excitation for the 10-story structure. The maximum displacement for the 10th story is 0.4101 m and the optimum TMD is effective in reducing it to 0.2622 with 36.06% performance. The time history plot for this excitation is also shown in Figure 7.8. The reduction of the maximum point and the effectiveness of the TMD on the damping of the vibrations are seen. Also, the maximum top-story acceleration is reduced to 12.0121 m/s^2 from 19.2864 m/s^2 for the same excitation.

Earthquake	Component	x_{10}		$\ddot{x}_{10} + \ddot{x}_g$	
		Without TMD	With TMD	Without TMD	With TMD
1	FN	0.3693	0.2382	15.8042	8.2525
	FP	0.3110	0.2713	12.9883	9.5841
2	FN	0.1326	0.1086	6.3276	5.0632
	FP	0.2236	0.1461	9.2066	6.4152
3	FN	0.2590	0.1569	12.7887	7.6020
	FP	0.4101	0.2622	19.2864	12.0021
4	FN	0.1118	0.1070	5.0418	4.2105
	FP	0.1317	0.1391	5.4565	4.4950
5	FN	0.1110	0.0652	5.3268	3.2047
	FP	0.1894	0.1082	7.8952	4.2661
6	FN	0.0765	0.0670	4.5812	3.7041
	FP	0.0705	0.0888	4.3957	4.4892
7	FN	0.1112	0.0999	5.9113	5.6091
	FP	0.1013	0.0882	5.1205	5.2906
8	FN	0.1045	0.1310	4.9963	5.3914
	FP	0.0764	0.0806	3.2676	3.0263
9	FN	0.1548	0.1150	8.4409	6.3647
	FP	0.2235	0.1903	9.8120	7.7423
10	FN	0.0407	0.0280	2.0715	1.6708
	FP	0.0396	0.0373	1.9932	1.2496
11	FN	0.1797	0.1371	7.4196	5.1637
	FP	0.1139	0.0815	4.9984	3.0643
12	FN	0.0834	0.0636	6.0349	3.5787
	FP	0.1369	0.1390	6.1439	5.7271
13	FN	0.1467	0.1539	8.9522	6.9083
	FP	0.0949	0.0950	5.0056	5.5921
14	FN	0.1139	0.0724	6.6829	6.0693
	FP	0.1223	0.1293	6.0778	5.5444
15	FN	0.1236	0.0919	6.0631	4.9267
	FP	0.1847	0.1375	9.9501	6.9884
16	FN	0.0848	0.1501	5.5291	5.7908
	FP	0.0837	0.0815	3.3533	3.4759

Earthquake	Component	x_{10}		$\ddot{x}_{10} + \ddot{x}_g$	
		Without TMD	With TMD	Without TMD	With TMD
17	FN	0.1151	0.0925	5.1103	4.4676
	FP	0.1375	0.0811	6.2135	4.8139
18	FN	0.1829	0.1426	8.5183	7.0506
	FP	0.1398	0.0946	7.7027	6.1423
19	FN	0.1608	0.1088	7.6721	5.2595
	FP	0.3547	0.1922	13.8343	8.5836
20	FN	0.1085	0.0704	6.6454	4.9365
	FP	0.1514	0.1169	7.1653	6.2917
21	FN	0.0851	0.0639	4.5123	3.1191
	FP	0.0614	0.0314	2.8126	1.5212
22	FN	0.0847	0.0606	5.3753	3.9533
	FP	0.1013	0.0756	5.2738	4.6214

Table 7.10. *Responses of the 10-story structure under far-field records*

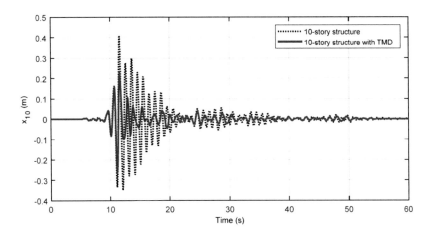

Figure 7.8. *Top-story displacement of the 10-story structure (DUZCE/BOL090)*

The maximum top-story displacement under near-field records with pulses is 0.6457 m and it is reduced to 0.5052 m by using the optimum TMD (21.76% reduction). The critical excitation, which is the FN component

(Northhr/Rrs-032) of 01 Rinaldi Receiving Sta. of the 1994 Northridge earthquake, was used to obtain the top-story displacement time history plot, given in Figure 7.9.

Earthquake	Component	x_{10}		$\ddot{x}_{10} + \ddot{x}_g$	
		Without TMD	With TMD	Without TMD	With TMD
1	FN	0.2159	0.1281	9.4237	5.3776
	FP	0.1509	0.1503	6.6648	5.8373
2	FN	0.2254	0.1712	9.5061	7.4246
	FP	0.2302	0.2110	9.5597	8.2970
3	FN	0.1045	0.1033	4.6150	3.5535
	FP	0.1546	0.1196	6.9255	4.2733
4	FN	0.3562	0.2679	13.5591	8.8910
	FP	0.1702	0.1438	7.3747	6.3768
5	FN	0.1538	0.1424	7.0769	5.8079
	FP	0.1281	0.0758	7.1845	4.6857
6	FN	0.1957	0.1820	11.1409	8.1833
	FP	0.2837	0.2301	11.7546	8.7199
7	FN	0.2135	0.1744	11.3048	9.3217
	FP	0.3367	0.2761	15.4409	12.0723
8	FN	0.1372	0.1889	6.3171	6.8368
	FP	0.1086	0.0696	5.1209	3.8508
9	FN	0.6457	0.5052	27.3301	17.4081
	FP	0.2700	0.2159	13.4820	9.2885
10	FN	0.2137	0.2090	9.2710	7.3446
	FP	0.2716	0.2935	14.8096	13.4099
11	FN	0.1105	0.0746	5.8749	2.8830
	FP	0.1141	0.0528	5.0043	1.9691
12	FN	0.4373	0.2742	18.2119	12.2201
	FP	0.4048	0.2349	17.9800	9.6555
13	FN	0.1912	0.1823	8.0242	5.8896
	FP	0.2293	0.1285	8.9202	4.4755
14	FN	0.1925	0.1176	8.0411	6.1547
	FP	0.2878	0.1706	11.5440	7.4123

Table 7.11. *Responses of the 10-story structure under near-field records with pulses*

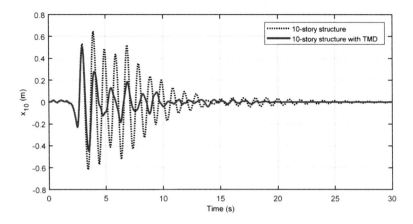

Figure 7.9. *Top-story displacement of the 10-story structure (Northhr/Rrs-032)*

Earthquake	Component	x_{10}		$\ddot{x}_{10} + \ddot{x}_g$	
		Without TMD	With TMD	Without TMD	With TMD
1	FN	0.2725	0.1793	11.0867	6.8370
	FP	0.1759	0.1165	8.7232	5.6536
2	FN	0.1551	0.1137	8.0885	6.3350
	FP	0.1668	0.1680	9.7206	9.3497
3	FN	0.0997	0.0703	3.9616	2.8200
	FP	0.1684	0.0945	6.1314	3.9323
4	FN	0.1485	0.1238	9.5579	7.9522
	FP	0.1907	0.0944	8.4289	5.5497
5	FN	0.0437	0.0456	2.8066	2.7346
	FP	0.0972	0.0776	4.6745	3.7971
6	FN	0.2075	0.1445	11.2136	7.9960
	FP	0.1304	0.1282	7.0685	6.6956
7	FN	0.1323	0.1198	9.9175	7.6878
	FP	0.1981	0.1421	9.3531	6.7038
8	FN	0.2499	0.2030	18.7204	15.8961
	FP	0.1048	0.1061	7.0698	6.4187
9	FN	0.3936	0.3095	17.5241	12.7042
	FP	0.2374	0.1441	11.7440	10.8559

Earthquake	Component	x_{10}		$\ddot{x}_{10} + \ddot{x}_g$	
		Without TMD	With TMD	Without TMD	With TMD
10	FN	0.1150	0.1138	5.9471	4.6714
	FP	0.3162	0.2385	13.9212	9.1484
11	FN	0.1307	0.1059	6.0181	4.1239
	FP	0.1228	0.1198	5.9340	4.6953
12	FN	0.3129	0.2043	12.9713	8.0932
	FP	0.2153	0.1699	9.5389	5.8795
13	FN	0.9920	0.5311	41.4756	20.8102
	FP	0.3063	0.1698	15.0598	7.6624
14	FN	0.3218	0.2761	11.7795	8.5293
	FP	0.1886	0.1509	7.7362	4.7067

Table 7.12. *Responses of the 10-story structure under near-field records without pulses*

The critical excitation of near-field records without pulses is FN component (Chichi/Tcu084-271) of the Tcu084 record of the 1999 Chi-Chi earthquake for the 10-story structure. As seen from the top-story displacement plot given in Figure 7.10, the optimum TMD is effective in reducing the peak value by 20.07% (from 0.9920 m to 0.5311 m).

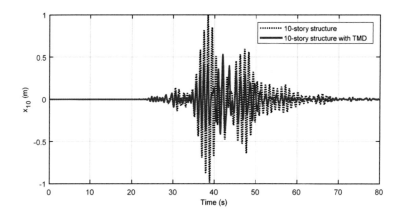

Figure 7.10. *Top-story displacement of the 10-story structure (Chichi/Tcu084-271)*

In the time domain, the maximum responses for the 40-story structure are presented in Tables 7.13–7.15. In these tables, the maximum results are given for both cases of maximum mass ratio.

Earthquake	Component	x_{10}			$\ddot{x}_{10} + \ddot{x}_g$		
		Without TMD	Case 1	Case 2	Without TMD	Case 1	Case 2
1	FN	0.2922	0.2850	0.2769	4.8623	4.7520	4.8420
	FP	0.2727	0.2233	0.2648	4.9423	5.5232	4.8834
2	FN	0.3287	0.2948	0.3210	3.3403	5.0772	3.2797
	FP	0.3486	0.2793	0.3291	3.1108	4.7877	3.0670
3	FN	0.4901	0.3382	0.4052	3.6592	8.0792	3.5815
	FP	0.2931	0.2347	0.2985	6.2482	6.7095	6.1251
4	FN	0.2378	0.2209	0.2160	2.0845	2.7074	2.0568
	FP	0.3234	0.2984	0.3022	3.0633	3.8086	3.0129
5	FN	0.5701	0.3705	0.3490	1.8986	3.2737	1.7188
	FP	0.4188	0.3296	0.3689	2.8242	3.5652	2.6286
6	FN	0.4795	0.3464	0.4319	2.2561	4.4601	2.1070
	FP	0.3469	0.2961	0.3254	2.4169	4.6619	2.3542
7	FN	0.2543	0.2088	0.2305	2.5383	6.2970	2.5014
	FP	0.2403	0.2396	0.2122	2.2089	5.8867	2.1374
8	FN	0.3128	0.2279	0.2910	3.1928	3.4189	3.0974
	FP	0.1981	0.1469	0.1832	1.9873	2.9447	1.9651
9	FN	1.4433	0.7533	1.0769	4.8147	4.0122	3.9386
	FP	0.5285	0.3417	0.4180	3.9598	5.0122	3.9028
10	FN	0.1658	0.1390	0.1601	0.8272	1.6813	0.8409
	FP	0.4850	0.5374	0.4486	1.5452	1.6854	1.4830
11	FN	0.4879	0.5153	0.4809	3.4140	3.6909	3.4239
	FP	0.4117	0.3764	0.4138	2.1439	2.2846	2.2017

Earthquake	Component	x_{10}			$\ddot{x}_{10} + \ddot{x}_g$		
		Without TMD	Case 1	Case 2	Without TMD	Case 1	Case 2
12	FN	0.2436	0.2131	0.2367	1.7939	2.8637	1.8231
	FP	0.2219	0.2454	0.2386	4.3720	5.0739	4.2680
13	FN	0.1943	0.1670	0.1957	3.9979	5.5859	3.9189
	FP	0.1924	0.1266	0.1527	2.6245	4.9397	2.6643
14	FN	0.2853	0.2253	0.2514	2.0954	6.0325	2.0191
	FP	0.4002	0.3758	0.3963	2.8984	5.4905	2.8606
15	FN	0.5659	0.3977	0.5061	3.4267	6.1291	3.2162
	FP	0.5621	0.4635	0.5536	2.6492	6.6539	2.6272
16	FN	0.7282	0.2935	0.3850	3.5269	3.6304	3.5055
	FP	0.4586	0.3432	0.3600	1.9760	3.2566	1.9468
17	FN	0.3750	0.2234	0.2624	2.8304	4.9268	2.7687
	FP	0.4061	0.3018	0.3248	2.3584	3.7651	1.8918
18	FN	0.2185	0.2147	0.2212	3.4061	4.3505	3.3848
	FP	0.2076	0.1580	0.1982	2.5629	4.9132	2.5141
19	FN	1.6377	1.0625	1.2877	5.0558	4.1746	4.1287
	FP	1.9278	1.6892	1.7826	6.1731	4.5408	5.3554
20	FN	0.2429	0.2974	0.2406	2.6781	5.0635	2.6104
	FP	0.2295	0.1410	0.1490	2.9085	5.0203	2.8554
21	FN	0.5883	0.3387	0.4751	2.4535	2.6689	2.3626
	FP	0.2370	0.1686	0.2077	0.9904	1.7330	0.8970
22	FN	0.1024	0.0867	0.0917	1.3802	3.5177	1.3512
	FP	0.1448	0.1085	0.1332	2.2727	3.6330	2.2450

Table 7.13. *Responses of the 40-story structure under far-field records*

Earthquake	Component	x_{10}			$\ddot{x}_{10} + \ddot{x}_g$		
		Without TMD	Case 1	Case 2	Without TMD	Case 1	Case 2
1	FN	0.9555	0.8055	0.9165	4.3428	4.4556	4.1246
	FP	2.2888	1.5169	2.0299	6.8914	6.7629	6.0929
2	FN	0.6599	0.5527	0.5456	3.4359	4.2673	3.4530
	FP	1.7659	1.2614	1.4549	6.2306	8.8257	6.1116
3	FN	0.4391	0.2740	0.3896	2.4019	3.5059	2.3471
	FP	1.0167	0.8291	0.9613	3.7016	4.4648	3.5401
4	FN	1.1657	1.1077	1.1677	7.6609	8.9766	7.5722
	FP	0.8347	0.4181	0.5809	4.0297	4.4268	4.0058
5	FN	0.3623	0.3246	0.3532	3.0572	5.6333	2.9996
	FP	0.7790	0.6418	0.7443	3.0652	3.8672	2.8766
6	FN	0.8675	0.5804	0.7257	4.5306	6.0259	4.4733
	FP	0.7798	0.7134	0.7649	6.4902	7.1655	6.3749
7	FN	0.3352	0.2914	0.3121	4.1974	7.4609	4.1180
	FP	0.7505	0.6846	0.7344	6.4979	9.9769	6.3617
8	FN	1.9216	1.4268	1.6975	6.8344	10.9192	6.5257
	FP	0.5554	0.3752	0.4861	2.3450	7.8364	2.1513
9	FN	0.8639	0.6846	0.7921	11.6282	9.9869	11.5128
	FP	0.6187	0.5221	0.5758	5.1647	6.9083	5.0746
10	FN	0.6237	0.4558	0.5736	5.0746	6.9267	4.8771
	FP	0.9390	0.7677	0.8926	7.0513	9.1464	6.9089
11	FN	0.5183	0.3610	0.4722	2.3327	3.0393	2.2875
	FP	0.4472	0.2514	0.3094	1.4601	2.2604	1.4277
12	FN	3.1361	2.0005	2.3028	12.1423	7.6718	8.6289
	FP	1.1577	0.9672	1.0853	6.7557	7.9618	6.4150
13	FN	1.2386	1.2029	1.2214	6.1225	5.9385	5.1657
	FP	1.1097	1.1599	1.0408	4.1691	4.0931	3.1410
14	FN	1.4232	0.8479	0.8584	3.9849	3.9349	3.8082
	FP	1.2290	1.0120	1.1217	5.5273	5.6801	5.2564

Table 7.14. *Responses of the 40-story structure under near-field records with pulses*

Earthquake	Component	x_{10}			$\ddot{x}_{10} + \ddot{x}_g$		
		Without TMD	Case 1	Case 2	Without TMD	Case 1	Case 2
1	FN	1.2436	0.7617	1.0809	5.9727	8.5253	5.3713
	FP	1.0734	0.7335	0.8328	5.0065	9.0017	4.6325
2	FN	0.3417	0.2807	0.3242	3.3732	8.3351	3.2609
	FP	0.2496	0.2587	0.2150	3.8054	8.9856	3.8781
3	FN	0.2163	0.2012	0.2310	2.1930	3.6929	2.1407
	FP	0.4396	0.2224	0.3066	2.6648	2.8891	2.6714
4	FN	0.4329	0.2336	0.3209	2.7785	9.7955	2.4568
	FP	0.5405	0.3151	0.3910	2.4301	10.4285	2.3156
5	FN	0.1445	0.1249	0.1400	1.5664	4.7302	1.5489
	FP	0.1456	0.1286	0.1420	2.1042	3.8375	2.0714
6	FN	0.3740	0.2523	0.3354	3.9188	5.8252	3.7466
	FP	0.1864	0.1655	0.1824	3.3813	5.6537	3.3437
7	FN	0.2390	0.1728	0.2222	3.4209	6.2938	3.3707
	FP	0.3442	0.2244	0.2941	4.6077	6.4762	4.5045
8	FN	1.0199	0.7391	0.8844	5.7668	14.3103	5.6493
	FP	0.3606	0.3127	0.3409	2.7603	10.7399	2.7214
9	FN	0.3659	0.2609	0.3373	7.4937	10.3620	7.4208
	FP	0.6782	0.4640	0.5951	4.8309	12.2230	4.4784
10	FN	0.3083	0.2383	0.2536	3.5648	3.7057	3.4094
	FP	0.6040	0.4731	0.5051	4.6685	7.1175	4.6441
11	FN	1.9469	1.2295	1.5422	6.8494	5.7939	5.3973
	FP	1.5798	1.0581	1.2584	6.5296	5.9817	5.7488
12	FN	0.7826	0.7768	0.7425	4.7846	5.7389	4.7023
	FP	1.1129	0.5987	0.6873	4.0426	5.3779	3.9305
13	FN	1.0034	0.8266	0.8666	10.7592	14.6794	11.3775
	FP	0.6722	0.4286	0.5607	4.0907	4.4096	4.0988
14	FN	1.2876	1.1295	1.2428	6.3268	5.7601	6.0415
	FP	1.1516	0.7893	0.9258	4.4826	3.7376	3.8755

Table 7.15. *Responses of the 40-story structure*
under near-field records without pulses

In far-field records, the FP component (Chichi/Chy101-N) of the CHY101 record of the 1999 Chi-Chi earthquake is the critical one for the 40-story structure and the optimum TMD is effective in reducing the maximum displacement of the 40th story from 1.9278 m to 1.6892 m for case 1 and to 1.7826 m for case 2. The response is plotted in Figure 7.11, and the effectiveness of the optimum TMD obtaining a rapid steady-state response is seen; it is distinct in case 1 which has a big optimum mass.

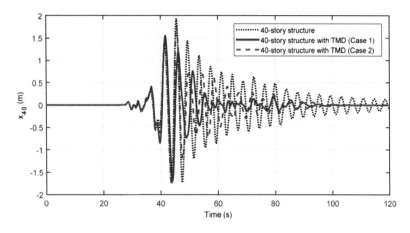

Figure 7.11. *Top-story displacement of the 40-story structure (Chichi/Chy101-N). For a color version of this figure, see www.iste.co.uk/toklu/metaheuristics.zip*

The history plot given in Figure 7.12 is the displacement of the 40th story of the structure under the FN component (Chichi/Tcu065-272) of the Tcu065 record of the 1999 Chi-Chi earthquake and it is the critical one of the near-field records with pulses. The optimum TMD provides a 36.21% reduction in case 1 and a 26.59% reduction in case 2.

The maximum top-story displacement of a 40-story structure is 1.9469 m for near-field records without pulses. This value belongs to the analyses of the FN component (Kocaeli/Ypt-180) of the Yarımca record of the 1999 Kocaeli earthquake and is plotted in Figure 7.13. The maximum displacement, which is 1.9469 m, is reduced to 1.2295 m and to 1.5422 m for cases 1 and 2, respectively.

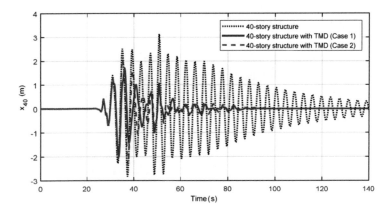

Figure 7.12. *Top-story displacement of the 40-story structure (Chichi/Tcu065-272). For a color version of this figure, see www.iste.co.uk/toklu/metaheuristics.zip*

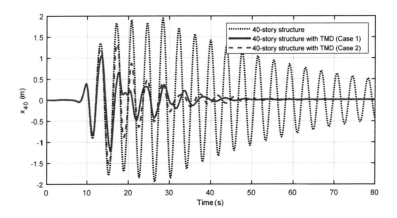

Figure 7.13. *Top-story displacement of the 40-story structure (Kocaeli/Ypt-180). For a color version of this figure, see www.iste.co.uk/toklu/metaheuristics.zip*

7.2. Optimum design of base isolation systems

Base isolation systems are generally used as support for critical structures in earthquake zones. The critical structures must not be damaged by strong ground motions in order to service people after the earthquakes. Also, the vibration-sensitive contents of structures are protected. In this case, the objective of base isolation systems is to protect structural integrity by reducing inter-story drifts and to reduce total floor acceleration to protect the vibration-sensitive contents, such as historical valuables, valuable equipment and devices affected by high acceleration (Nagarajaiah and Xiaohong 2000).

A base isolated structure contains three main parts as follows:

– an isolation system including bearings and dampers;

– the superstructure, which is an ordinary structure including structural members like beams, columns and floors;

– a rigid base floor between the superstructure and isolation system.

The motion of a shear building supported by a base isolation system (Figure 7.14) is formulated with the equation of motion, as seen in equation [7.21] (Matsagar and VJangid 2004).

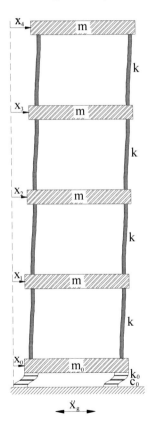

Figure 7.14. *Schematics of the prototype seismically isolated structure*

$$M\ddot{x}(t)+C\dot{x}(t)+Kx(t)=-MI(\ddot{x}_0+\ddot{x}_g)$$ [7.21]

The matrices and vectors in equation [7.21] are similar to the equations presented in section 7.1.2. The only differences are the elimination of the values corresponding to the TMD. The relative acceleration of the base is represented by $\ddot{x}_0$ and the equation of the motion of the base is formulated as equation [7.22].

$$m_0\ddot{x}_0+c_1\dot{x}_1+k_1x_1+f_r=-m_0\ddot{x}_g$$ [7.22]

The mass of a rigid base floor is shown with m_0; the parameters of the base isolation system are the effective stiffness (k_0) and effective viscous damping coefficient (c_0). The restoring force (f_r) is formulated as equation [7.23] if a Kelvin-Voight viscoelastic model is considered for the isolation system. The Kelvin–Voight viscoelastic model is a rheological model containing a linear spring and a linear viscous damper connected in parallel.

$$f_r=k_0x_0+c_0\dot{x}_0$$ [7.23]

The isolation period (T_0) and the isolation damping (ξ_0), which are the design variables of an optimum design process, are used to find the following parameters:

$$k_0=\frac{4M\pi^2}{T_0^2}$$ [7.24]

$$c_0=\frac{4M\pi\xi_0}{T_0}$$ [7.25]

M is the total mass of the structure and a rigid base floor.

The optimum design of base isolation systems has been investigated by using several metaheuristic algorithms such as Genetic Algorithms (GA) (Pourzeynali and Zarif 2008) and Harmony Search (HS) (Nigdeli et al. 2013).

Nigdeli et al. (2013) employed HS to minimize the ratio of peak roof acceleration (PRA) and peak ground acceleration (PGA), as given in equation [7.26]. This value must be lower than the desired percentage (DP) defined by the users of the developed optimization code:

$$\frac{PRA}{PGA} \le DP \qquad\qquad [7.26]$$

The DP value may also be defined as zero. The value of DP iteratively increased after several iterations. In this case, the physically possible minimum value can be found. Also, the optimization of the T_0 and ξ_0 values is done by considering a design constraint (equation [7.27]):

$$\max |x_0| \le x_{max} \qquad\qquad [7.27]$$

The peak isolation system displacement (x_0) must be lower than the limit displacement set by the user (x_{max}).

Case	Isolation system displacement limit (cm)	Isolation system damping ratio limit (%)	Optimum isolation system stiffness (kN)	Optimum isolation system damping (kNs/m)	Optimum isolation period (s)	Optimum isolation damping ratio (%)
1	40	30	30.308	7.283	2.564	29.98
2	40	40	22.675	8.458	2.950	39.98
3	40	50	18.590	9.353	3.259	48.77
4	50	30	19.869	5.892	3.152	29.73
5	50	40	16.241	7.105	3.486	39.61
6	50	50	14.493	8.367	3.961	49.35
7	60	30	15.232	5.192	3.600	29.88
8	60	40	13.564	6.505	3.815	39.65
9	60	50	12.500	7.661	3.974	48.63

Table 7.16. *Optimum parameters of the isolation system (Nigdeli et al. 2013)*

In the numerical example in Nigdeli *et al.* (2013), a four-story structure was adjusted to obtain a fundamental fixed-base period of 0.45 s (Alhan and Surmeli 2011). The optimum isolation system parameters were found according to several cases of isolation system displacement and damping ratio limit, as seen in Table 7.16. The cases and target period range of the isolation system (2 s<T_0<4 s) were chosen according to Pan *et al.* (2005).

The optimum results were found using data analyses of six earthquakes, downloaded from the Pacific Earthquake Engineering Research Center (PEER) database. The information from the records is given in Table 7.17 and includes PGA, velocity (PGV) and displacement (PGD) values.

Earthquake	Date	Station	Component	PGA (g)	PGV (cm/s)	PGD (cm)
Imperial Valley	1940	117 El Centro Array #9	I-ELC180	0.313	29.8	13.32
Kern Country	1952	1095 Taft Lincoln School	TAFT111	0.178	17.5	8.99
Tabas	1978	9101 Tabas	TAB-TR	0.852	121.4	94.58
Loma Prieta	1989	16 LGPC	LGP000	0.563	94.8	41.18
Erzincan	1992	95 Erzincan	ERZ-NS	0.515	83.9	27.35
Northridge	1994	24514 Sylmar Olive View Med FF	SYL360	0.843	129.6	32.68

Table 7.17. *Earthquake data used in the optimization of a seismic isolation system (Nigdeli et al. 2013)*

The maximum responses of the optimum base-isolated structure are given in Table 7.18 for optimization earthquakes.

As a conclusion to the study of Nigdeli *et al.* (2013), the optimization of base isolation systems is effective in the reduction of story drift ratios. Thus, the second-order effects are reduced. Also, optimization under several excitations is very useful in the optimization because the critical excitations are different (Loma Prieta in cases 1–4 and Tabas for cases 5–9) between cases. Also, the base isolation system displacement is the maximum for other excitations (Erzincan in cases 1–3 and Loma Prieta in cases 4–9).

Case	Response	Earthquake record					
		Taft	El Centro	Erzincan	Tabas	Sylmar	Loma Prieta
1	x_0 (cm)	5.334	10.462	33.976	32.171	35.923	39.941
	PRA/PGA	0.258	0.288	0.503	0.309	0.330	0.501
2	x_0 (cm)	5.159	8.173	28.398	35.282	31.461	39.987
	PRA/PGA	0.230	0.256	0.419	0.289	0.313	0.432
3	x_0 (cm)	4.964	7.698	26.695	35.971	27.985	38.214
	PRA/PGA	0.219	0.249	0.396	0.281	0.310	0.394
4	x_0 (cm)	5.828	10.446	33.040	43.433	37.297	49.937
	PRA/PGA	0.208	0.211	0.359	0.257	0.263	0.448
5	x_0 (cm)	5.474	9.414	29.966	49.523	31.493	45.398
	PRA/PGA	0.197	0.205	0.336	0.254	0.265	0.387
6	x_0 (cm)	5.130	8.458	27.588	49.695	28.913	40.731
	PRA/PGA	0.198	0.216	0.345	0.256	0.278	0.356
7	x_0 (cm)	5.983	11.201	33.391	59.262	35.441	53.523
	PRA/PGA	0.176	0.166	0.293	0.255	0.227	0.397
8	x_0 (cm)	5.561	9.814	30.401	59.565	31.416	46.914
	PRA/PGA	0.179	0.180	0.298	0.255	0.242	0.358
9	x_0 (cm)	5.228	8.869	28.222	56.411	29.704	42.140
	PRA/PGA	0.184	0.195	0.313	0.252	0.257	0.337

Table 7.18. *Response of optimum base-isolated structure (Nigdeli et al. 2013)*

8

Applications of Metaheuristic Algorithms to Structural Analysis

Metaheuristic algorithms are, by definition, applicable to problems that are formulated as optimization problems. Structural problems can be defined as optimization ones in two different ways. One of the two ways is by optimizing the topology, shape and weight of a structure, leaving the analysis part to an appropriate method, finite element method (FEM), for instance. This optimization technique is very common and is used by quite a large number of researchers. The other optimization is on the analysis part of the problem. This approach, called Total Potential Optimization using Metaheuristic Algorithms (TPO/MAs), is more recent than the other, still in its infancy, but seems very promising given the success of the applications carried out until now. As the name implies, this technique is nothing but the application of the fundamental minimum energy principle to structures.

8.1. Fundamentals of the method

As stated in any treatise on energy principles about structures, any body assumes a shape corresponding to the minimum potential energy configuration under the effect of the loads acting on it (see, for instance, Oden (1967), Tauchert (1974) and Reddy (2017)). This means that the deflected shape of a body is the one with the minimum potential energy; any other configuration will have a larger potential energy. Thus, if the total potential energy of a structure is written in terms of deflection parameters, and if a minimization process is carried out to determine the deflections resulting in the minimum total potential energy, then the deflections of the structure can be determined, corresponding to the stable equilibrium configuration. After determination of the deflections in a structure, all other unknowns, related to strains, stresses, member forces, shear forces, moments, etc. can be found using the relevant relations between them. This method is a displacement method, as compared to the force method, where the basic unknowns are the forces and not the displacements.

The basic equations related to this method are the following:

$$\Pi = S - W \tag{8.1}$$

$$S = \int_V \sigma\varepsilon \, d\varepsilon \tag{8.2}$$

$$W = \sum_i P_i \delta_i, \ i \epsilon I_P \tag{8.3}$$

where:

– Π is the total potential energy in the structure;

– S is the strain energy in the system;

– W is the work done by the generalized external forces;

– V is the total volume of the structure;

– σ and ε are the stress and strain at a point in the structure, respectively;

– P_i's are the external generalized forces and δ_i's are the corresponding generalized displacements;

– I_P is the set covering all the external generalized forces.

In the above equations, δ_i, the generalized deflections corresponding to generalized loads, and ε, the generalized strains in the structure, are directly related to the displacements x_i, with $i \epsilon I_P$ defining the considered configuration of the system. On the other hand, the stresses σ are also functions of x_i's through the constitutive equation of the material of the type $\sigma = \sigma(\varepsilon)$.

The procedure proposed can therefore be written in the form

$$\text{Min } \Pi(\mathbf{x}) = S(\mathbf{x}) - W(\mathbf{x}), \text{ such that } \mathbf{g}(\mathbf{x}) \leq \mathbf{0}, \mathbf{h}(\mathbf{x}) = \mathbf{0} \tag{8.4}$$

where $\mathbf{g}(\mathbf{x})$ and $\mathbf{h}(\mathbf{x})$ are, respectively, the inequality and equality constraints about the state variables $\mathbf{x}$, the vector containing the displacements x_i defining the deflected configuration of the structure.

For structural problems that are not large and have no complexities, equation 8.4 can be applied directly to the system to find the deflected shape. However, in general, the technique that is followed is dividing the structure into finite elements and writing $S(\mathbf{x})$ in equation [8.4] for each element separately; followed by summing all total potentials to reach the general total potential of the system:

$$\min \Pi = \sum_{j=1}^{J} S_j(\mathbf{x}) - W(\mathbf{x}) \tag{8.5}$$

where J is the total number of finite elements in the structure and S_j is the strain energy in element j.

We can apply an appropriate optimization technique in the minimization necessary at equation [8.5]. Until now, different metaheuristic algorithms, and some hybrid ones, have been used for this purpose, and thus the method is called Total Potential Optimization using Metaheuristic Algorithms (TPO/MA). This method could also be called Finite Element Method with Energy Minimization (FEMEM).

The difference between TPO/MA and FEM can be clarified here. In FEM, after dividing the structure into finite parts, equilibrium equations are written in the matrix form for each element; these matrices are then combined to reach a general matrix equation of the form $\mathbf{Kx} = \mathbf{p}$, where $\mathbf{K}$ is the stiffness matrix of the system, $\mathbf{p}$ is the vector containing the acting loads and $\mathbf{x}$ is the vector of state variables, i.e. the displacements that are being searched for. Thus, for a problem with n state variables, in FEM, a matrix equation will be solved when the matrix is nxn, and in TPO/MA, a functional will be minimized with respect to n variables. We can note that both methods have their advantages and disadvantages, as will be seen in the examples given in the following paragraphs. The main disadvantages of these methods can be described as:

– In FEM, the matrix can be ill-conditioned in some problems, and for nonlinear problems, the matrix is not an invariable of the problem. Indeed, for nonlinear problems, **K** becomes a function of the applied loads and the displacements, thus the solution necessitates iterations and results in a loss of accuracy and more CPU time. Additionally, if there is more than one equilibrium configuration of a system, it is impossible to solve the problem through a normal application of FEM.

– In TPO/MA, as n gets bigger, accurately finding n variables by minimizing a single functional becomes more difficult and time-consuming.

Problem type	FEM	TPO/MA
Linear systems (small)	– Very efficient	– Efficient but takes more time than FEM
Linear systems (large)	– Very efficient	– Accuracy may not be at desired level, particularly for displacements with small contributions to total potential energy
Nonlinear systems	– Requires an expert for formulation and application – Probable loss of accuracy due to accumulated errors	– Solving is as easy as a linear problem – No special attention – No accumulated errors
Multiple solutions	– Impossible to solve directly	– Efficient
Under-constrained systems	– Impossible to solve – Matrix ill-conditioned	– Solving is as easy as a well-constrained problem – No special attention
Unilateral constraints	– Not possible to solve directly – Difficult application of iterations with continuous user intervention	– Solving is as easy as a well-constrained problem – Make an arrangement about limits of the deflections in the question
Degenerate problems	– Impossible to solve – Matrix ill-conditioned	– Solving is as easy as a non-degenerate problem – No special attention

Table 8.1. *Comparison between FEM and TPO/MA*

A comparison between FEM and TPO/MA is shown in Table 8.1 concerning the static analysis of structures. Historically, TPO/MA was first introduced in 2004 (Toklu 2004a, 2004b), though it was known by another name. This method has since been applied to analyses of trusses and truss-like structures, such as tensegrity structures, cable nets and plates, as explained in the following sections.

8.2. Applications to structures, generalities

Structures are analyzed through TPO/MA following the steps below:

– structures are divided into finite elements;

– the displacements at nodes, **x**, which are at the edges of the elements, are considered as unknowns of the problem;

– strain energies in finite elements are written in terms of these unknowns;

– the total strain energy of the system, S(**x**), is written as the sum of these energies;

– work done by the external forces W(**x**) is also written in terms of the unknowns **x** and then;

– **x** is searched for using an appropriate optimization algorithm that makes the total potential, Π(**x**), a minimum, where

$$\Pi(\mathbf{x}) = S(\mathbf{x}) - W(\mathbf{x}) \tag{8.6}$$

8.3. Applications to trusses and truss-like structures

In solving trusses and truss-like structures, members, which may be prismatic bars or cables forming the system, are taken as the finite elements, and the joint displacements, forming a vector **x**, are taken as the unknowns of the problem. In this case, the strains can be written in terms of a candidate **x** (Toklu 2004a, 2004b), and the strain energy density in member j can be written as in equation [8.7]:

$$e_j(\varepsilon) = \int_0^{\varepsilon_j} \sigma(\varepsilon)\, d\varepsilon \tag{8.7}$$

where ε_j is the strain in member j. For linear elastic materials, the integral in equation [8.7] can be readily taken as $e_j = \sigma(\varepsilon_j)\,\varepsilon_j\,/2$. Furthermore, for these materials, if $\sigma = E\varepsilon$, with E as the modulus of elasticity, the strain energy density in member j becomes $e_j = E\,\varepsilon_j^2\,/2$.

For nonlinear materials, the stress–strain relationship may be given as a collection of piecewise continuous lines, as in Figure 8.1. In this case, the strain energy density e will be calculated as the area A, shown in Figure 8.1 as the sum of the areas of trapezoids (Toklu and Uzun 2016). In some other cases, stress–strain relationships may be defined as functions, as in equation [8.8], from which the strain energy density can be calculated, as in equation [8.9] (Toklu 2004b):

$$\sigma = (a_1\varepsilon + a_2)^{a_3} + a_4\varepsilon + a_5, \quad a_3 > 0 \tag{8.8}$$

$$e = \frac{(a_1\varepsilon + a_2)^{a_3+1}}{a_1(a_3 + 1)} + \frac{a_4}{2}\varepsilon^2 + a_5\varepsilon. \tag{8.9}$$

Computation of strain energy densities will be followed by the computation of strain energies in each member by multiplying the densities by the volumes of the elements, and the sum of these will give the total strain energy S of the system. Writing the work done by external forces in terms of the same displacement vector **x** enables us to form the total potential Π of the system after equation [8.6]. The problem is then to calculate Π for a series of candidates of **x**, to determine the one making Π a minimum.

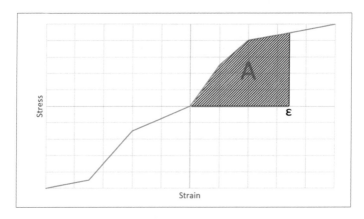

Figure 8.1. *Stress–strain diagram as a collection of piece-wise continuous lines*

In the TPO/MA method, as the name implies, the optimization methods used are metaheuristic and hybrid ones. In the applications performed, several metaheuristic methods are tried, including simulated annealing, harmony search, the ant colony optimization, teaching–learning-based optimization, the flower pollination algorithm and the Jaya algorithm.

With the properties described above, problems with nonlinear materials can be solved with no special attention, after, of course, defining the stress–strain relation and its integral at any given level of strain. Geometric nonlinearities, i.e. large deformations, also do not pose any problem in the applications, since, regardless of the deflections, the strains become calculated accordingly, and there are no accumulated errors that we encounter in FEM applications, when a technique like the P–δ method (Gaiotti and Smith 1989) or a similar one is used. In progressive failure analyses, we may arrive at a point where, due to the failure of one or more members, the truss may become a degenerate one with an ill-conditioned stiffness matrix, in classical terms. In these cases, FEM becomes unusable since the relevant matrix equation becomes unsolvable. These degenerate structures also pose no problem to TPO/MA, since there is always an equilibrium position for a system, regardless of its stiffness matrix. The figures below give some examples of truss behaviors that cannot be solved by FEM.

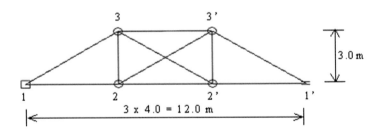

Figure 8.2. *Ten-member truss, geometry with the numbering of joints (Toklu 2004c)*

The method, as presented here, is applied to several problems concerning trusses (Toklu 2004a, 2004b; Toklu and Toklu 2013; Toklu *et al.* 2013, 2015a; Bekdaş *et al.* 2019a). Some other applications are on trusses with unilateral boundary conditions (Temür *et al.* 2014), those with under-constrained structures (Toklu *et al.* 2017) and trusses under thermal effects (Toklu *et al.* 2015b). Truss-like structures like cables (Kayabekir *et al.* 2018b; Bekdaş *et al.* 2019b) and tensegric structures (Toklu and Uzun 2016) are also studied with TPO/MA. Toklu and Arditi (2014) studied smart trusses for space applications.

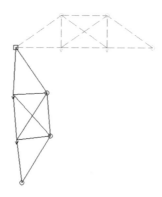

Figure 8.3. *Deflected shape of the 10-bar truss when the right support is removed (loading 2 x 1,000 kN at the lower chord, at joints 2 and 2')*

a) P = 30 kN   b) P = 40 kN

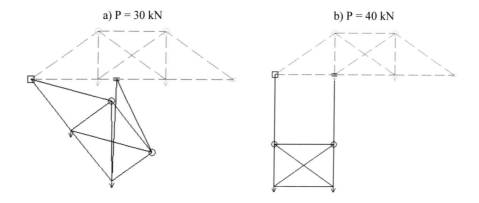

Figure 8.4. *Deflected shapes of the 10-member truss after yielding. Material: elastic–plastic*

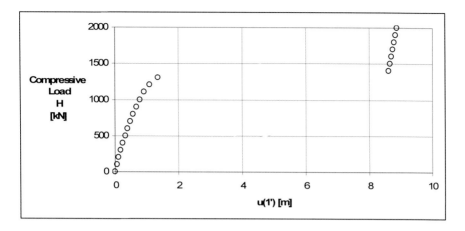

Figure 8.5. *Load–deformation curve for the 10-member truss. Load H applied at joint 1'. Deformations measured at the same point. Note the snap-through behavior between 1,300 and 1,400 kN*

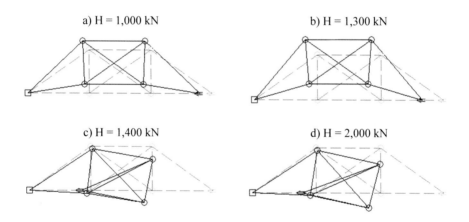

Figure 8.6. *Deflected shapes for the 10-member truss under the effect of load H applied at joint 1'*

8.4. Applications to plates

Plates are analyzed through TPO/MA following the same principles applied to trusses and truss-like structures, dividing them into finite elements and considering the displacements, **x**, at nodes as the unknowns. The following steps are the same as in trusses:

– strain energies in finite elements are written in terms of these unknowns;

– the total strain energy of the system, S(**x**), is written as the sum of these energies;

– work done by the external forces W(**x**) is also written in terms of the unknowns **x**; and then

– **x** is searched for through an appropriate optimization algorithm that makes the total potential, $\Pi(\mathbf{x}) = S(\mathbf{x}) - W(\mathbf{x})$, a minimum.

The finite elements can be chosen from many possibilities, as in normal FEM. For the moment, the choice for these introductory applications is the triangular element shown in Figure 8.7.

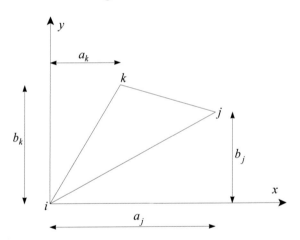

Figure 8.7. *The triangular element*

Applications performed until now have been on plane stress systems for linear constitutive equations (Kayabekir *et al.* 2020b) and nonlinear constitutive equations (Toklu *et al.* 2020). It should be noted that, because of the lack of any previous study, the nonlinear stress–strain equations employed in these studies are hypothetical.

An example in these studies can be given from the applications on plane stress analyses of thick-walled pipe with internal pressure. This system is analyzed by the well-known linear stress–strain relation and with three hypothetical nonlinear ones:

$$\sigma_x = \frac{E}{1-v^2}(\varepsilon_x + v\varepsilon_y) \qquad [8.10a]$$

$$\sigma_y = \frac{E}{1-v^2}(\varepsilon_y + v\varepsilon_x) \qquad [8.10b]$$

$$\tau = \frac{E}{1-v^2}(\frac{1-v}{2}\gamma) \qquad [8.10c]$$

Case 1: Nonlinear stress–strain relation 1

$$\sigma_x = \frac{E}{1-v^2}\mathrm{sgn}(\varepsilon_x + v\varepsilon_y)(\varepsilon_x + v\varepsilon_y)^2 \qquad [8.11a]$$

$$\sigma_y = \frac{E}{1-v^2}\mathrm{sgn}(\varepsilon_y + v\varepsilon_x)(\varepsilon_y + v\varepsilon_x)^2 \qquad [8.11b]$$

$$\tau = \frac{E}{1-v^2}\mathrm{sgn}(\frac{1-v}{2}\gamma)(\frac{1-v}{2}\gamma)^2 \qquad [8.11c]$$

Case 2: Nonlinear stress–strain relation 2

$$\sigma_x = \frac{E}{1-v^2}(\varepsilon_x + v\varepsilon_y)^3 \qquad [8.12a]$$

$$\sigma_y = \frac{E}{1-v^2}(\varepsilon_y + v\varepsilon_x)^3 \qquad [8.12b]$$

$$\tau = \frac{E}{1-v^2}(\frac{1-v}{2}\gamma)^3 \qquad [8.12c]$$

Case 3: Nonlinear stress–strain relation 3

$$\left.\begin{array}{l} \sigma_x = \dfrac{E}{1-v^2}(\varepsilon_x + v\varepsilon_y) \\[4pt] \sigma_y = \dfrac{E}{1-v^2}(\varepsilon_y + v\varepsilon_x) \\[4pt] \tau = \dfrac{E}{1-v^2}(\dfrac{1-v}{2}\gamma) \end{array}\right\} \ for \ \varepsilon_x < \varepsilon_y, \ E = E_1 \ and \ for \ \varepsilon_x > \varepsilon_y, \ E = E_2 \qquad [8.13]$$

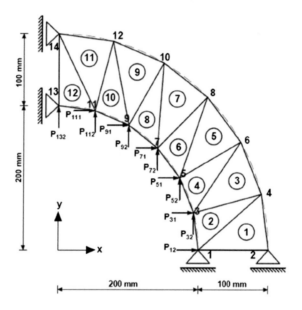

Figure 8.8. *Twelve-member, 14-node model of a quarter of a thick-walled pipe with internal pressure. Plane stress example*

Because of the lack of any study with nonlinear stress–strain relations, the results obtained for this problem can be compared only to the linear case, where the exact answer for the shortening of the radius is 0.55 mm for the data given. This value is found to be between 0.4718 and 0.47933 in FEM applications with different meshing options, and 0.4959 and 0.5058 in corresponding TPO/MA applications (Toklu *et al.* 2020), which is much better than FEM.

The cited studies will be continued using applications on plane strain and plane bending problems, with more complicated finite elements.

8.5. Further studies on the analysis of structures with TPO/MA

It can be said that analysis of trusses and truss-like structures using TPO/MA is highly developed, although there are still problems to address in order to show the real power of the method. Applications on plates are only at an introductory level. There is still almost an infinite number of structures like shells, beams, frames, volumes, partly or fully infinite structures and every type of combination of them, under very different loading conditions with all types of constraints. When we consider that FEM of today is the output of thousands of studies carried out until now, we can see that TPO/MA, or FEMEM in another way of saying, needs much more research in the coming years, to arrive at the level it deserves.

Future Trends

With the advances observed in the speed and memory capacity of computers, and also in the theoretical and practical knowledge about coding and metaheuristics, it seems like there will be much progress in the application of MAs to structural design and analysis.

It is obvious that the level that we are currently at is not sufficient for analyzing and designing structural systems using MAs. In the analysis, the level attained enables engineers to perform the static analysis of trusses if the number of unknowns is reasonable, say less than 1,000. For linear systems and some nonlinear problems, the TPO/MA method has many disadvantages compared to FEM and similar techniques, especially regarding CPU time and accuracy. On the other hand, TPO/MA, also called FEMEM, is advantageous to classical FEM in many nonlinear cases, where the stress–strain relation is nonlinear and/or the deflections are large, but especially for degenerate systems that have missing or failed members and supports, under-constrained systems and those with unilateral supports. The fact that TPO/MA is the only approach used when solving these systems means that more research has to be carried out to improve this method. The expected research will certainly concentrate on applications on structures other than trusses and truss-like ones, continuing work done on plates and dealing with beams, frames, shells, volumes, infinite structures, etc., and any combination of them. Other research directions will be on diminishing runtime, diminishing the number and effect of parameters of the algorithm used and producing user-friendly software that can be run by non-specialized engineers for any type of structure.

Coming to design, again, the current level is far from satisfactory. As stated in the preceding chapters, the existing commercial packages for design optimization are all based on mathematical optimization techniques other than MAs. Up until today, MAs have mostly been applied on trusses. There are some applications on frames, grids, etc. but not a satisfactory number and they are not easy to use. The main research direction in design will probably be on extending this procedure to more general structures and not only on dimension or weight optimization, but also on shape and topology optimization. As stated above, the current level in design optimization using MAs is far from sufficient. For instance, in the size optimization of trusses, which is very primitive compared to shape and topology optimization, the search for profiles with automatic grouping is still in its infancy. The main aim of designing a civil engineering complex structure like a dam, a suspension bridge or a high-rise building completely with metaheuristic algorithms is still far beyond the scope of engineers and researchers.

The final problem is to make the design of a civil engineering edifice, simultaneously with the analysis, using metaheuristic algorithms. This simply involves combining the design, $\min(F_1(\mathbf{x_1}))$, where F_1 is a function giving the weight, or, more generally, the cost of the structure, and the analysis, $\min(F_2(\mathbf{x_2}))$, where F_2 is the function giving the total potential energy of the system as defined in TPO/MA, under the constraints of the problem, with respect to relevant parameters $\mathbf{x_1}$ and $\mathbf{x_2}$. In this case, the problem becomes a multi-objective optimization, one to be solved using equality and inequality constraints. This procedure is yet to be applied.

Artificial intelligence (AI) has been defined in many different ways, changing with accomplishments in the area of research. The design and analysis of structures by way of MAs can be considered to be simple applications of AI, if we consider that they involve gradual learning by a machine (in this case, a computer), using the rule of thumb. The simultaneous design and analysis of structures using MAs will be a real and difficult application of AI.

There is a trend among AI researchers to say that if a target is achieved, it is no longer an application of AI, and the real goal is still one step ahead. At this point, we can say that there are further applications of AI in this area, like the simultaneous design and analysis of structures using MAs, involving all types of optimization, such as topology, shape and size. A giant step forward would be combining the results of this procedure with all of the other activities of construction, including material and manpower procuration, site choice and organization, planning of works from start to finish, making financial analyses and making relevant decisions as data continues to flow from construction activities, from the financial side, from weather conditions, etc. If we consider that such problems may be encountered in the future, while performing operations on remote space objects with important time lags in communications, we would be apt to conclude that this work is very difficult and needs many more decades to be succeed, using knowledge and sources of AI in combination with Building Information Modeling (BIM).

References

ACI (2005). 318M-05: Building code requirements for structural concrete and commentary. American Concrete Institute, Michigan, USA.

Addis, B. (2003). Inventing a history for structural engineering design. *Proceedings of the First International Congress on Construction History*, 20, 24.

Ahrari, A. and Atai, A.A. (2013). Fully stressed design evolution strategy for shape and size optimization of truss structures. *Computers & Structures*, 123, 58–67.

AISC (2010). 360-10: Specification for structural steel buildings. American Institute of Steel Construction, Chicago, USA.

Akin, A. and Saka, M.P. (2010). Optimum detailed design of reinforced concrete continuous beams using the harmony search algorithm. *Proceedings of the Tenth International Conference on Computational Structures Technology*, Stirlingshire, UK.

Akin, A. and Saka, M.P. (2012). Optimum detailing design of reinforced concrete plane frames to ACI 318-05 using the harmony search algorithm. In *Proceedings of the Eleventh International Conference on Computational Structures Technology*, Topping, B.H.V. (ed.). Civil-Comp Press, Stirlingshire.

Akin, A. and Saka, M.P. (2015). Harmony search algorithm based optimum detailed design of reinforced concrete plane frames subject to ACI 318-05 provisions. *Computers & Structures*, 147, 79–95.

Alhan, C. and Surmeli, M. (2011). Shear building representations of seismically isolated buildings. *Bulletin of Earthquake Engineering*, 9, 1643–1671.

Almufti, S.M. (2019). Historical survey on metaheuristics algorithms. *International Journal of Scientific World*, 7(1), 1.

Beer, F.P., Johnston, R., Dewolf, J., Mazurek, D. (2006). *Mechanics of Materials.* McGraw-Hill, New York.

Bekdaş, G. (2014). Optimum design of axially symmetric cylindrical reinforced concrete walls. *Structural Engineering and Mechanics*, 51(3), 361–375.

Bekdaş, G. (2015). Harmony search algorithm approach for optimum design of post-tensioned axially symmetric cylindrical reinforced concrete walls. *Journal of Optimization Theory and Applications*, 164(1), 342–358.

Bekdaş, G. and Nigdeli, S.M. (2011). Estimating optimum parameters of tuned mass dampers using harmony search. *Engineering Structures*, 33(9), 2716–2723.

Bekdaş, G. and Nigdeli, S.M. (2012). Cost optimization of t-shaped reinforced concrete beams under flexural effect according to ACI 318. *3rd European Conference of Civil Engineering*, Paris, France.

Bekdaş, G. and Nigdeli, S.M. (2015). Optimum design of reinforced concrete beams using teaching learning based optimization. *3rd International Conference on Optimization Techniques in Engineering*, Rome, Italy.

Bekdaş, G. and Nigdeli, S.M. (2016a). Optimum design of reinforced concrete columns employing teaching learning based optimization. *12th International Congress on Advances in Civil Engineering*, Istanbul, Turkey.

Bekdaş, G. and Nigdeli, S.M.T. (2016b). Bat algorithm for optimization of reinforced concrete columns. *Joint Annual Meeting of GAMM and DMV*, Braunschweig, Germany.

Bekdaş, G. and Nigdeli, S.M. (2017a). Modified harmony search for optimization of reinforced concrete frames. In *Harmony Search Algorithm. Advances in Intelligent Systems and Computing*, Del Ser, J. (ed.). Springer-Verlag, Berlin, Heidelberg.

Bekdaş, G. and Nigdeli, S.M. (2017b). Metaheuristic based optimization of tuned mass dampers under earthquake excitation by considering soil – Structure interaction. *Soil Dynamics and Earthquake Engineering*, 92, 443–461.

Bekdaş, G. and Nigdeli, S.M. (2017c). Optimum reduction of flexural effect of axially symmetric cylindrical walls with post-tensioning forces. *KSCE Journal of Civil Engineering*, 22, 2425–2432.

Bekdaş, G., Nigdeli, S.M., Yang, X.-S. (2014). Metaheuristic optimization for the design of reinforced concrete beams under flexure moments. *5th European Conference of Civil Engineering*, Florence, Italy.

Bekdaş, G., Nigdeli, S.M., Yang, X.S. (2015). Sizing optimization of truss structures using flower pollination algorithm. *Applied Soft Computing*, 37, 322–331.

Bekdaş, G., Nigdeli, S.M., Aydn, A. (2017a). Optimization of tuned mass damper for multi-story structures by using impulsive motions. *2nd International Conference on Civil and Environmental Engineering*, Cappadocia, Turkey.

Bekdaş, G., Nigdeli, S.M., Kayabekir, A.E. (2017b). Optimization of spans of multi-story frames using teaching learning based optimization. *3rd International Conference on Engineering and Natural Sciences*, Budapest, Hungary.

Bekdaş, G., Kayabekir, A.E., Nigdeli, S.M., Toklu, Y.C. (2019a). Advanced energy based analyses of trusses employing hybrid metaheuristics. *The Structural Design of Tall and Special Buildings*, 28(9), e1609.

Bekdaş, G., Nigdeli, S.M., Toklu, Y.C. (2019b). Total potential energy minimization using metaheuristic algorithms for spatial cable systems with increasing second order effects. *12th HSTAM International Congress on Mechanics*, Thessaloniki, Greece.

Bendsøe, M.P. and Sigmund, O. (2004). *Topology Optimization: Theory, Methods and Applications*. Springer-Verlag, Berlin, Heidelberg.

Benvenuto, E. (1990). *An Introduction to the History of Structural Mechanics*. Springer-Verlag, New York.

Bishop, R.E.D. and Welbourn, D.B. (1952). The problem of the dynamic vibration absorber. *Engineering (London)*, 4, 174–769.

Boussaï, D.I., Lepagnot, J., Siarry, P. (2013). A survey on optimization metaheuristics. *Information Sciences*, 237, 82–117.

Çakmak, A.Ş., Taylor, R.M., Durukal, E. (2009). The structural configuration of the first dome of Justinian's Hagia Sophia (AD 537–558): An investigation based on structural and literary analysis. *Soil Dynamics and Earthquake Engineering*, 29(4), 693–698.

Camp, C.V. (2007). Design of space trusses using big bang–big crunch optimization. *Journal of Structural Engineering*, 133(7), 999–1008.

Camp, C.V. and Assadollahi, A. (2013). CO_2 and cost optimization of reinforced concrete footings using a hybrid big bang-big crunch algorithm. *Structural and Multidisciplinary Optimization*, 48(2), 411–426.

Camp, C.V. and Bichon, B.J. (2004). Design of space trusses using ant colony optimization. *Journal of Structural Engineering*, 130(5), 741–751.

Camp, C.V. and Farshchin, M. (2014). Design of space trusses using modified teaching learning based optimization. *Engineering Structures*, 62–63, 87–97.

Camp, C.V. and Huq, F. (2013). CO_2 and cost optimization of reinforced concrete frames using a big bang-big crunch algorithm. *Engineering Structures*, 48, 363–372.

Camp, C.V., Pezeshk, S., Hansson, H. (2003). Flexural design of reinforced concrete frames using a genetic algorithm. *Journal of Structural Engineering*, 129, 105–111.

Chen, Y., Feng, J., Wu, Y. (2012). Novel form-finding of tensegrity structures using ant colony systems. *Journal of Mechanisms and Robotics*, 4(3), 031001.

Chickermane, H. and Gea, H.C. (1996). Structural optimization using a new local approximation method. *International Journal for Numerical Methods in Engineering*, 39, 829–846.

Choi, W.H., Kim, J.M., Park, G.J. (2016). Comparison study of some commercial structural optimization software systems. *Structural and Multidisciplinary Optimization*, 54(3), 685–699.

Coello, C.A. and Christiansen, A.D. (2000). Multiobjective optimization of trusses using genetic algorithms. *Computers & Structures*, 75(6), 647–660.

Coello, C.C., Hernandez, F.S., Ferrera, F.A. (1997). Optimal design of reinforced concrete beams using genetic algorithms. *Expert Systems with Applications*, 12, 101–108.

Collette, Y. and Siarry, P. (2013). *Multiobjective Optimization: Principles and Case Studies*. Springer Science & Business Media, Berlin.

Connelly, R. and Terrell, M. (1995). Globally rigid symmetric tensegrities. *Structural Topology*, 21, 59–78.

Copeland, B.J. (2000). What is artificial intelligence? [Online]. Available at: http://www.alanturing.net/turing_archive/pages/Reference%20Articles/What%20is%20AI.html.

Cormen, T.H., Leiserson, C.E., Rivest, R.L., Stein, C. (2009). *Introduction to Algorithms*. MIT Press, Cambridge.

Cowan, H.J. (1977). *A Historical Outline of Architectural Science*, 2nd edition. Applied Science Publishers, London.

Cross, H. (1930). *Selected Papers in Arches, Continuous Frames, Frames and Conduits*. University of Illinois Press, Champaign.

Das, S. and Suganthan, P.N. (2011). Differential evolution: A survey of the state-of-the-art. *IEEE Transactions on Evolutionary Computation*, 15(1), 4–31.

Dede, T. and Ayvaz, Y. (2015). Combined size and shape optimization of structures with a new meta-heuristic algorithm. *Applied Soft Computing*, 28, 250–258.

Degertekin, S.O. and Hayalioglu, M.S. (2013). Sizing truss structures using teaching learning-based optimization. *Computers & Structures*, 119, 177–188.

Den Hartog, J.P. (1947). *Mechanical Vibrations*. McGraw-Hill, New York.

Desu, N.B., Deb, S.K., Dutta, A. (2006). Coupled tuned mass dampers for control of coupled vibrations in asymmetric buildings. *Structural Control and Health Monitoring*, 13, 897–916.

Do, D.T., Lee, S., Lee, J. (2016). A modified differential evolution algorithm for tensegrity structures. *Composite Structures*, 158, 11–19.

Dorigo, M., Maniezzo, V., Colorni, A. (1996). The ant system: Optimization by a colony of cooperating agents. *IEEE Transactions on Systems, Man and Cybernetics, Part B*, 26, 29–41.

Eaton, L.K. (2001). Hardy cross and the "Moment distribution method". *Nexus Network Journal*, 3(2), 15–24.

Engau, A. and Sigler, D. (2020). Pareto solutions in multicriteria optimization under uncertainty. *European Journal of Operational Research*, 281(2), 357–368.

Estrada, G.G., Bungartz, H.J., Mohrdieck, C. (2006). Numerical form-finding of tensegrity structures. *International Journal of Solids and Structures*, 43, 6855–6868.

European Committee for Standardization (2004). Eurocode 2: Design of Concrete Structures, Brussels, Belgium.

European Committee for Standardization (2005). Eurocode 3: Design of Steel Structures, Brussels, Belgium.

Falcon, K.C., Stone, B.J., Simcock, W.D., Andrew, C. (1967). Optimization of vibration absorbers: A graphical method for use on idealized systems with restricted damping. *Journal Mechanical Engineering Science*, 9, 374381.

Farshidianfar, A. and Soheili, S. (2013a). Ant colony optimization of tuned mass dampers for earthquake oscillations of high-rise structures including soil–structure interaction. *Soil Dynamics and Earthquake Engineering*, 51, 14–22.

Farshidianfar, A. and Soheili, S. (2013b). ABC optimization of TMD parameters for tall buildings with soil structure interaction. *Interaction and Multiscale Mechanics*, 6, 339–356.

Fedghouche, F. and Tiliouine, B. (2012). Minimum cost design of reinforced concrete T-beams at ultimate loads using Eurocode2. *Engineering Structures*, 42, 43–50.

Felippa, C.A. (2001). A historical outline of matrix structural analysis: A play in three acts. *Computers & Structures*, 79(14), 1313–1324.

FEMA (2009). Quantification of Building Seismic Performance Factors, FEMA P-695. Applied Technology Council for the Federal Emergency Management Agency, Washington, D.C., USA.

Fletcher, R. (2013). *Practical Methods of Optimization*. John Wiley & Sons, New York.

Fleury, C. and Braibant, V. (1986). Structural optimization: A new dual method using mixed variables. *International Journal for Numerical Methods in Engineering*, 23, 409–428

Frahm, H. (1911). Device for damping of bodies. United States Patent 0989958.

Fuller, R.B. (1962). Tensile-integrity structures. United States Patent 3063521.

Gaiotti, R. and Smith, B.S. (1989). P-Delta analysis of building structures. *Journal of Structural Engineering*, 115(4), 755–770.

Gandibleux, X., Sevaux, M., Sörensen, K., T'Kindt, V. (2004). *Metaheuristics for Multiobjective Optimisation*. Springer Science & Business Media, Berlin.

Gandomi, A.H. and Alavi, A.H. (2012). Krill herd: A new bio-inspired optimization algorithm. *Communications in Nonlinear Science and Numerical Simulation*, 17(12), 4831–4845.

Gandomi, A.H., Talatahari, S., Yang, X.S., Deb, S. (2013a). Design optimization of truss structures using cuckoo search algorithm. *The Structural Design of Tall and Special Buildings*, 22(17), 1330–1349.

Gandomi, A.H., Yang, X.S., Alavi, A.H. (2013b). Cuckoo search algorithm: A metaheuristic approach to solve structural optimization problems. *Engineering with Computers*, 29, 17–35.

Geem, Z.W., Kim, J.H., Loganathan, G.V. (2001). A new heuristic optimization algorithm: Harmony search. *Simulation*, 76, 60–68.

Glover, F. (1986). Future paths for integer programming and links to artificial intelligence. *Computers & Operations Research*, 13(5), 533–549.

Gold, S. and Krishnamurty, S. (1997). Trade-offs in robust engineering design. *Proceedings of the 1997 ASME Design Engineering Technical Conferences*, Saramento, California.

Goldberg, D.E. and Samtani, M.P. (1986). Engineering optimization via genetic algorithm. *Proceedings of the Ninth Conference on Electronic Computation*. ASCE, New York, 471–482.

Govindaraj, V. and Ramasamy, J.V. (2005). Optimum detailed design of reinforced concrete continuous beams using genetic algorithms. *Computers & Structures*, 84, 34–48.

Govindaraj, V. and Ramasamy, J.V. (2007). Optimum detailed design of reinforced concrete frames using genetic algorithms. *Engineering Optimization*, 39(4), 471–494.

Hadi, M.N.S. and Arfiadi, Y. (1998). Optimum design of absorber for MDOF structures. *Journal of Structural Engineering*, 124, 12721280.

Hao, J.K. and Solnon, C. (2020). *A Guided Tour of Artificial Intelligence Research 2: AI Algorithms*. Springer, Cham.

Hasançebi, O. and Azad, S.K. (2014). Discrete size optimization of steel trusses using a refined big bang–big crunch algorithm. *Engineering Optimization*, 46(1), 61–83.

Hasançebi, O. and Azad, S.K. (2015). Adaptive dimensional search: A new metaheuristic algorithm for discrete truss sizing optimization. *Computers & Structures*, 154, 1–16.

Hetenyi, M. (1946). *Beams on Elastic Foundation*. The University of Michigan Press, Ann Arbor.

Holland, J.H. (1975). *Adaptation in Natural and Artificial Systems*. University of Michigan Press, Ann Arbor.

Hsu, Y.L. and Liu, T.C. (2007). Developing a fuzzy proportional derivative controller optimization engine for engineering design optimization problems. *Engineering Optimization*, 39(6), 679–700.

Ioi, T. and Ikeda, K. (1978). On the dynamic vibration damped absorber of the vibration system. *Bulletin of the JSME*, 21, 6471.

Jakiela, M.J., Chapman, C., Duda, J., Adewuya, A., Saitou, K. (2000). Continuum structural topology design with genetic algorithms. *Computer Methods in Applied Mechanics and Engineering*, 186(2), 339–356.

Karaboga, D. (2005). An idea based on honey bee swarm for numerical optimization. Technical Report, Erciyes University, Turkey.

Karaboga, D. and Basturk, B. (2008). On the performance of artificial bee colony (ABC) algorithm. *Applied Soft Computing*, 8(1), 687–697.

Kaveh, A. and Sabzi, O. (2011). A comparative study of two meta-heuristic algorithms for optimum design of reinforced concrete frames. *International Journal of Civil Engineering*, 9(3), 193–206.

Kaveh, A. and Talatahari, S. (2009a). Particle swarm optimizer, ant colony strategy and harmony search scheme hybridized for optimization of truss structures. *Computers & Structures*, 87(5), 267–283.

Kaveh, A. and Talatahari, S. (2009b). Size optimization of space trusses using big bang–big crunch algorithm. *Computers & Structures*, 87(17), 1129–1140.

Kayabekir, A.E., Nigdeli, S.M., Bekdaş, G. (2017). Optimization of spans of frame structures employing teaching learning based optimization. *Proceedings of the 2nd International Conference on Civil and Environmental Engineering*, Cappadocia, Turkey.

Kayabekir, A.C., Nigdeli, S.M., Bekdaş, G., Toklu, Y.C. (2018a). The analyses of spatial cable structures employing new generation metaheuristic methods. *Proceedings of the 9th GRACM 2018 International Congress on Computational Mechanics*, Chania, Greece.

Kayabekir, A.E., Bekdaş, G., Nigdeli, S.M., Yang, X.S. (2018b). A comprehensive review of the flower pollination algorithm for solving engineering problems. In *Nature-Inspired Algorithms and Applied Optimization*, Yang, X.S. (ed.). Springer, Cham.

Kayabekir, A.E., Toklu, Y.C., Bekdaş, G., Nigdeli, S.M., Yücel, M., Geem, Z.W. (2020). A novel hybrid harmony search approach for the analysis of plane stress systems via total potential optimization. *Applied Sciences*, 10(7), 2301.

Kelesoglu, O. (2007). Fuzzy multiobjective optimization of truss-structures using genetic algorithm. *Advances in Engineering Software*, 38(10), 717–721.

Kennedy, J. and Eberhart, R.C. (1995). Particle swarm optimization. *Proceedings of the IEEE International Conference on Neural Networks No. IV*, 1942–1948, Perth, Australia.

Khajehzadeh, M., Taha, M.R., El-Shafie, A., Eslami, M. (2011). Modified particle swarm optimization for optimum design of spread footing and retaining wall. *Journal of Zhejiang University SCIENCE A*, 12(6), 415–427.

Khajehzadeh, M., Taha, M.R., El-Shafie, A., Eslami, M. (2012). Optimization of shallow foundation using gravitational search algorithm. *Journal of Applied Engineering and Technology*, 4(9), 1124–1130.

Khajehzadeh, M., Taha, M.R., Eslami, M. (2013). A new hybrid firefly algorithm for foundation optimization. *National Academy Science Letters*, 36(3), 279–288.

Kicinger, R., Arciszewski, T., De Jong, K. (2005). Evolutionary computation and structural design: A survey of the state-of-the-art. *Computers & Structures*, 83(23), 1943–1978.

Kirkpatrick, S., Gelatt, C.D., Vecchi, M.P. (1983). Optimization by simulated annealing. *Science*, 220(4598), 671–680.

Koohestani, K. (2012). Form-finding of tensegrity structures via genetic algorithm. *International Journal of Solids and Structures*, 49(5), 739–747.

Koumousis, V.K. and Arsenis, S.J. (1998). Genetic algorithms in optimal detailed design of reinforced concrete members. *Computer-Aided Civil and Infrastructure Engineering*, 13, 43–52.

Koumousis, V.K. and Georgiou, P.G. (1994). Genetic algorithms in discrete optimization of steel truss roofs. *Journal of Computing in Civil Engineering*, 8(3), 309–325.

Krenk, S. and Høgsberg, J. (2013). *Statics and Mechanics of Structures*. Springer Science & Business Media, Berlin.

Kress, G. and Keller, D. (2007). Script for the lecture in structural optimization. Swiss Federal Institute of Technology Zurich, Centre of Structure Technologies, Zurich.

Kurrer, K.E. (2018). *The History of the Theory of Structures: Searching for Equilibrium*. John Wiley & Sons, New York.

Lee, C. and Ahn, J. (2003). Flexural design of reinforced concrete frames by genetic algorithm. *Journal of Structural Engineering*, 129(6), 762–774.

Lee, K.S. and Geem, Z.W. (2004). A new structural optimization method based on the harmony search algorithm. *Computers & Structures*, 82(9), 781–798.

Leps, M. and Sejnoha, M. (2003). New approach to optimization of reinforced concrete beams. *Computers & Structures*, 81, 1957–1966.

Leung, A.Y.T. and Zhang, H. (2009). Particle swarm optimization of tuned mass dampers. *Engineering Structures*, 31, 715–728.

Leung, A.Y.T., Zhang, H., Cheng, C.C., Lee, Y.Y. (2008). Particle swarm optimization of TMD by non-stationary base excitation during earthquake. *Earthquake Engineering and Structural Dynamics*, 37, 1223–1246.

Li, J.P. (2015). Truss topology optimization using an improved species-conserving genetic algorithm. *Engineering Optimization*, 47(1), 107–128.

Li, J.P., Balazs, M.E., Parks, G.T. (2007a). Engineering design optimization using species-conserving genetic algorithms. *Engineering Optimization*, 39(2), 147–161.

Li, L.J., Huang, Z.B., Liu, F., Wu, Q.H. (2007b). A heuristic particle swarm optimizer for optimization of pin connected structures. *Computers & Structures*, 85(7), 340–349.

Li, Y., Feng, X.Q., Cao, Y.P., Gao, H. (2010). A Monte Carlo form-finding method for large scale regular and irregular tensegrity structures. *International Journal of Solids and Structures*, 47, 1888–1898.

Liu, M.Y., Chiang, W.L., Hwang, J.H., Chu, C.R. (2008). Wind-induced vibration of high-rise building with tuned mass damper including soil–structure interaction. *Journal of Wind Engineering and Industrial Aerodynamics*, 96(6), 1092–1102.

Lu, P., Chen, S., Zheng, Y. (2012). Artificial intelligence in civil engineering. Mathematical Problems in Engineering. Article ID 145974.

Mainstone, R.J. (1988). *Haghia Sophia: Architecture, Structure and Liturgy of Justinian's Great Church*. Thames and Hudson, London.

Mainstone, R.J. (1997). Structural analysis, structural insights, and historical interpretation. *The Journal of the Society of Architectural Historians*, 56(3), 316–340.

Majid, K.I. (1974). *Optimum Design of Structures*. Newnes-Butterworth, London.

Makris N. (1997). Rigidity-plasticity-viscosity: Can electrorheological dampers protect base-isolated structures from near-source ground motions. *Earthquake Engineering and Structural Dynamics*, 26, 571–591.

Manjarres, D., Landa-Torres, I., Gil-Lopez, S., Del Ser, J., Bilbao, M.N., Salcedo-Sanz, S., Geem, Z.W. (2013). A survey on applications of the harmony search algorithm. *Engineering Applications of Artificial Intelligence*, 26(8), 1818–1831.

Marano, G.C., Greco, R., Chiaia, B. (2010). A comparison between different optimization criteria for tuned mass dampers design. *Journal of Sound and Vibration*, 329, 4880–4890.

Masic, M., Skelton, R.E., Gill, P.E. (2005). Algebraic tensegrity form-finding. *International Journal of Solids and Structures*, 42, 4833–4858.

MathWorks, Inc. (2010). MATLAB R2010a. Natick, USA.

Matsagar, V.A. and Jangid, R.S. (2004). Influence of isolator characteristics on the response of base-isolated structures. *Earthquake Engineering and Structural Dynamics*, 11, 43–55.

Miguel, L.F.F., Lopez, R.H., Miguel, L.F.F. (2013). Multimodal size, shape, and topology optimisation of truss structures using the firefly algorithm. *Advances in Engineering Software*, 56, 23–37.

Miyamoto, H.K., Gilani, A.S.J., Gndodu, Y.Z. (2011). Innovative seismic retrofit of an iconic building. *Seventh National Conference on Earthquake Engineering*, Istanbul, Turkey.

Motro, R. (1984). Forms and forces in tensegrity systems. In *Proceedings of the Third International Conference on Space Structures*, Nooshin, H. (ed.). Elsevier, Amsterdam.

Motro, R. (1992). Tensegrity systems: The state of the art. *International Journal of Space Structures*, 7, 75–82.

Nagarajaiah, S. and Xiaohong, S. (2000). Response of base-isolated USC hospital building in Northridge earthquake. *Journal of Structural Engineering*, 126, 1177–1186.

Nigdeli, S.M. and Bekdaş, G. (2015). Teaching-learning based optimization for estimating tuned mass damper parameters. *3rd International Conference on Optimization Techniques in Engineering*, Rome, Italy.

Nigdeli, S.M. and Bekdaş, G. (2016). Detailed optimum design of reinforced concrete frame structures. *WSEAS Transactions on Applied and Theoretical Mechanics*, 11(28), 220–228.

Nigdeli, S.M. and Bekdaş, G. (2017). Optimum tuned mass damper design in frequency domain for structures. *KSCE Journal of Civil Engineering*, 21(3), 912–922.

Nigdeli, S.M., Bekdaş, G, Alhan, C. (2013). Optimization of seismic isolation systems via harmony search. *Engineering Optimization*, 46(11), 1553–1569.

Nigdeli, S.M., Bekdaş, G., Yang, X.-S. (2016). Application of the flower pollination algorithm in structural engineering. In *Metaheuristics and Optimization in Civil Engineering*, Yang, X.-S., Bekdaş, G., Nigdeli, S.M. (eds). Springer, Switzerland.

Nigdeli, S.M., Bekdaş, G., Aydn, A. (2017a). Metaheuristic based optimization of tuned mass dampers on single degree of freedom structures subjected to near fault vibrations. *Proceedings of the International Conference on Engineering and Natural Sciences*, Budapest, Hungary.

Nigdeli, S.M., Bekdaş, G., Yang, X.-S. (2017b). Optimum tuning of mass dampers by using a hybrid method using harmony search and flower pollination algorithm. In *Harmony Search Algorithm. Advances in Intelligent Systems and Computing*, Del Ser, J. (ed.). Springer-Verlag, Berlin, Heidelberg.

Nigdeli, S.M., Bekdaş, G., Toklu, Y.C. (2019). Total potential energy minimization using metaheuristic algorithms for spatial cable systems with increasing second order effects. *Proceedings of the 12th International Congress on Mechanics*, 22–25.

Nocedal, J. and Wright, S. (2006). *Numerical Optimization*. Springer Science & Business Media, Berlin.

Nowcki, H. (1974). Optimization in pre-contract ship design. In *Computer Applications in the Automation of Shipyard Operation and Ship Design*, Fujita, Y., Lind, K., Williams, T.J. (eds). Elsevier, New York.

Oden, J.T. (1967). *Mechanics of Elastic Structures*. McGraw-Hill, New York.

Ormondroyd, J. and Den Hartog, J.P. (1928). The theory of dynamic vibration absorber. *Journal of Applied Mechanics*, 50, 9–22.

Ozkul, T.A. and Kuribayashi, E. (2007). Structural characteristics of Hagia Sophia: I – A finite element formulation for static analysis. *Building and Environment*, 42(3), 1212–1218.

Pagitz, M. and Mirats Tur, J.M. (2009). Finite element based form-finding algorithm for tensegrity structures. *International Journal of Solids and Structures*, 46, 3235–3240.

Pan, P., Zamfirescu, D., Nakashima, M., Nakayasu, N., Kashiwa, H. (2005). Base-isolation design practice in Japan: Introduction to the post-Kobe approach. *Journal of Earthquake Engineering*, 9, 147–171.

Park, Y.C., Chang, M.H., Lee, T.Y. (2007). A new deterministic global optimization method for general twice differentiable constrained nonlinear programming problems. *Engineering Optimization*, 39(4), 397–411.

Parpinelli, R.S. and Lopes, H.S. (2011). New inspirations in swarm intelligence: A survey. *International Journal of Bio-Inspired Computation*, 3(1), 1–16.

Paya, I., Yepes, V., Gonzalez-Vidosa, F., Hospitaler, A. (2008). Multiobjective optimization of concrete frames by simulated annealing. *Computer-Aided Civil and Infrastructure Engineering*, 23, 596–610.

Paya-Zaforteza, I., Yepes, V., Hospitaler, A., Gonzalez-Vidosa, F. (2009). CO_2-optimization of reinforced concrete frames by simulated annealing. *Engineering Structures*, 31, 1501–1508

PEER (2005). The Pacific Earthquake Engineering Research Center: NGA database. University of California, Berkeley [Online]. Available at: http://peer.berkeley. edu/nga [Accessed November 2011].

Perez, R.E. and Behdinan, K. (2007). Particle swarm approach for structural design optimization. *Computers & Structures*, 85(19), 1579–1588.

Polya, G. (2004). *How to Solve It: A New Aspect of Mathematical Method*. Princeton University Press, Princeton.

Poulos, H.G. and Davis, E.H. (1974). *Elastic Solutions for Soil and Rock Mechanics*. John Wiley & Sons, New York.

Pourzeynali, S. and Zarif, M. (2008). Multi-objective optimization of seismically isolated high-rise building structures using genetic algorithms. *Journal of Sound and Vibration*, 311, 1141–1160.

Pourzeynali, S., Lavasani, H.H., Modarayi, A.H. (2007). Active control of high rise building structures using fuzzy logic and genetic algorithms. *Engineering Structures*, 29, 346–357.

Rajeev, S. and Krishnamoorthy, C.S. (1992). Discrete optimization of structures using genetic algorithms. *Journal of Structural Engineering*, 118(5), 1233–1250.

Rajeev, S. and Krishnamoorthy, C.S. (1998). Genetic algorithm-based methodology for design optimization of reinforced concrete frames. *Computer-Aided Civil and Infrastructure Engineering*, 13, 63–74.

Rao, S.S. (1996). *Engineering Optimization: Theory and Practice*, 3rd edition. John Wiley & Sons, Chichester.

Rao, R. (2016). Jaya: A simple and new optimization algorithm for solving constrained and unconstrained optimization problems. *International Journal of Industrial Engineering Computations*, 7(1), 19–34.

Rao, R.V., Savsani, V.J., Vakharia, D.P. (2011). Teaching–learning-based optimization: A novel method for constrained mechanical design optimization problems. *Computer-Aided Design*, 43(3), 303–315.

Ray, T. and Saini, P. (2001). Engineering design optimization using a swarm with an intelligent information sharing among individuals. *Engineering Optimization*, 33(6), 735–748.

Reddy, J.N. (2017). *Energy Principles and Variational Methods in Applied Mechanics*. John Wiley & Sons, New York.

Rieffel, J., Cuevas, F.V., Lipson, H. (2009). Automated discovery and optimization of large irregular tensegrity structures. *Computers & Structures*, 87, 368–379.

Sadek, F., Mohraz, B., Taylor, A.W., Chung, R.M. (1997). A method of estimating the parameters of tuned mass dampers for seismic applications. *Earthquake Engineering and Structural Dynamics*, 26, 617–635.

Sadollah, A., Bahreininejad, A., Eskandar, H., Hamdi, M. (2012). Mine blast algorithm for optimization of truss structures with discrete variables. *Computers & Structures*, 102, 49–63.

Saka, M.P., Hasançebi, O., Geem, Z.W. (2016). Metaheuristics in structural optimization and discussions on harmony search algorithm. *Swarm and Evolutionary Computation*, 28, 88–97.

Schmit, L.A. (1960). Structural design by systematic synthesis. *Proceedings of the Second National Conference on Electronic Computation*, Pittsburgh, USA.

Schutte, J.F. and Groenwold, A.A. (2003). Sizing design of truss structures using particle swarms. *Structural and Multidisciplinary Optimization*, 25(4), 261–269.

Šešok, D. and Belevičius, R. (2008). Global optimization of trusses with a modified genetic algorithm. *Journal of Civil Engineering and Management*, 4(3), 147–154.

Siddique, N. and Adeli, H. (2015). Applications of harmony search algorithms in engineering. *International Journal on Artificial Intelligence Tools*, 24(06), 1530002.

Simon, H.A. and Newell, A. (1958). Heuristic problem solving: The next advance in operations research. *Operations Research*, 6(1), 1–10.

Singh, M.P., Matheu, E.E., Suarez, L.E. (1997). Active and semi-active control of structures under seismic excitation. *Earthquake Engineering and Structural Dynamics*, 26, 193213.

Singh, M.P., Singh, S., Moreschi, L.M. (2002). Tuned mass dampers for response control of torsional buildings. *Earthquake Engineering and Structural Dynamics*, 31, 749–769.

Sirenko, S. (2009). Classification of heuristic methods in combinatorial optimization. *International Journal "Information Theories & Applications"*, 16(4), 303–322.

Snelson, K.D. (1965). Continuous tension, discontinuous compression structures. United States Patent 3169611.

Snowdon, J.C. (1959). Steady-state behavior of the dynamic absorber. *Journal of the Acoustical Society of America*, 31, 1096–1103.

Sonmez, M. (2011). Artificial bee colony algorithm for optimization of truss structures. *Applied Soft Computing*, 11(2), 2406–2418.

Sörensen, K. (2015). Metaheuristics – The metaphor exposed. *International Transactions in Operational Research*, 22(1), 3–18.

Sörensen, K., Sevaux, M., Glover, F. (2018). A history of metaheuristics [Online]. Available at: https://arxiv.org/pdf/1704.00853.pdf.

Srinivas, M. and Patnaik, L.M. (1994). Genetic algorithms: A survey. *Computer*, 27(6), 17–26.

Steinbuch, R. (2011). Bionic optimisation of the earthquake resistance of high buildings by tuned mass dampers. *Journal of Bionic Engineering*, 8, 335–344.

Steward, J.P., Chiou, S., Bray, J.D., Graves, R.W., Somerville, P.G., Abrahamson, N.A. (2001). Ground motion evalution procedures for performance-based design. Report, Pacific Earthquake Engineering Research Center, California.

Storn, R. and Price, K. (1997). Differential evolution – A simple and efficient heuristic for global optimization over continuous spaces. *Journal of Global Optimization*, 11(4), 341–359.

Suman, B. and Kumar, P. (2006). A survey of simulated annealing as a tool for single and multiobjective optimization. *Journal of the Operational Research Society*, 57(10), 1143–1160.

Talatahari, S. and Kaveh, A. (2015). Improved bat algorithm for optimum design of large-scale truss structures. *International Journal of Optimization in Civil Engineering*, 5(2), 241–254.

Tauchert, T.R. (1974). *Energy Principles in Structural Mechanics*. McGraw-Hill, New York.

Techasen, T., Wansasueb, K., Panagant, N., Pholdee, N., Bureerat, S. (2019). Simultaneous topology, shape, and size optimization of trusses, taking account of uncertainties using multi-objective evolutionary algorithms. *Engineering with Computers*, 35(2), 721–740.

Temur, R., Bekdaş, G., Turkan, Y.S., Toklu, Y.C. (2014). Investigating the behavior of the truss structures with unilateral boundary conditions. *Proceedings of the 5th European Conference of Civil Engineering*, Florence, Italy.

Tibert, A.G. and Pellegrino, S. (2003). Review of form finding methods for tensegrity structures. *International Journal of Space Structures*, 18(4), 209–223.

Timoshenko, S. (1983). *History of Strength of Materials: With a Brief Account of the History of Theory of Elasticity and Theory of Structures*. Dover Publications, New York.

Toğan, V. and Daloğlu, A.T. (2008). An improved genetic algorithm with initial population strategy and self-adaptive member grouping. *Computers & Structures*, 86(11), 1204–1218.

Toklu, Y.C. (2004a). Application of metaheuristic optimization techniques to structural analyses: Truss example. *Proceedings of the Structures 2004 Congress*, Nashville, Tennessee, USA.

Toklu, Y.C. (2004b). Nonlinear analysis of trusses through energy minimization. *Computers & Structures*, 82(20–21), 1581–1589.

Toklu, Y.C. (2004c). A new technique for nonlinear analysis of trusses. *Proceedings of the 6th International Congress on Advances in Civil Engineering*. Bogazici University, Istanbul, Turkey.

Toklu, Y.C. (2005). Aggregate blending using genetic algorithms. *Computer-Aided Civil and Infrastructure Engineering*, 20(6), 450–460.

Toklu, Y.C. (2009). Optimization in structural analysis and design. *Proceedings of the Structures Congress 2009: Don't Mess with Structural Engineers: Expanding Our Role*, 1–10. April 30–May 2, Austin, USA.

Toklu, Y.C. and Arditi, D. (2014). Smart trusses for space applications. *Proceedings of the 14th Biennial ASCE Conference on Engineering, Science, Construction, and Operations in Challenging Environments*, October 27–29, St. Louis, Missouri, USA.

Toklu, Y.C. and Bekdaş, G. (2014). Metaheuristics and engineering. *Proceedings of the 15th EU/ME Workshop*, Bilecik Şeyh Edebali University, Bilecik, Turkey.

Toklu, Y.C. and Toklu, N.E. (2013). Analysis of structures by total potential optimization using meta-heuristic algorithms (TPO/MA). In *Heuristics: Theory and Applications*, Siarry, P. (ed.). Nova Science, New York.

Toklu, Y.C. and Uzun, F. (2016). Analysis of tensegric structures by total potential optimization using metaheuristic algorithms. *Journal of Aerospace Engineering*, 29(5), 04016023.

Toklu, Y.C., Bekdaş, G., Temür, R. (2013). Analysis of trusses by total potential optimization method coupled with harmony search. *Structural Engineering and Mechanics*, 45(2), 183–199.

Toklu, Y.C., Temür, R., Bekdaş, G. (2015a). Computation of non-unique solutions for trusses undergoing large deflections. *International Journal of Computational Methods*, 12(3), 1550022.

Toklu, Y.C., Bekdaş, G., Temur, R. (2015b). Investigation of thermal effects on analyses of truss structures via metaheuristic approaches. *Proceedings of the 6th European Conference of Civil Engineering*, Budapest, Hungary.

Toklu, Y.C., Bekdaş, G., Nigdeli, S.M. (2017). Analysis of under-constrained and unstable structures by the method TPO/MA. *Proceedings of the 15th International Conference of Numerical Analysis and Applied Mathematics*, Thessaloniki, Greece.

Toklu, Y.C., Kayabekir, A.E., Bekdaş, G., Nigdeli, S.M., Yücel, M. (2020). Analysis of plane-stress systems via total potential optimization method considering nonlinear behavior. *Journal of Structural Engineering*, 146(11), 04020249.

Tsai, J. (2005). Global optimization of nonlinear fractional programming problems in engineering design. *Engineering Optimization*, 37(4), 399–409.

Wang, G.G. (2003). Adaptive response surface method using inherited Latin hypercube design points. *Transactions of the ASME*, 125, 210–220.

Warburton, G.B. (1982). Optimum absorber parameters for various combination of response and excitation parameters. *Earthquake Engineering and Structural Dynamics*, 10, 381–401.

Warburton, G.B. and Ayorinde, E.O. (1980). Optimum absorber parameters for simple systems. *Earthquake Engineering and Structural Dynamics*, 8, 197–217.

Weyland, D. (2010). A rigorous analysis of the harmony search algorithm: How the research community can be misled by a "novel" methodology. *International Journal of Applied Metaheuristic Computing*, 1(2), 50–60.

Whitman, R.V. and Richart, F.E. (1967). Design procedures for dynamically loaded foundations. *Journal of the Soil Mechanics and Foundations Division*, 93(6), 169.

Xu, X. and Luo, Y. (2010a). Force finding of tensegrity systems using simulated annealing algorithm. *Journal of Structural Engineering*, 136(8), 1027–1031.

Xu, X. and Luo, Y. (2010b). Form-finding of nonregular tensegrities using genetic algorithm. *Mechanics Research Communications*, 37, 85–91.

Yang, X.S. (2005). Engineering optimizations via nature-inspired virtual bee algorithms. *Lecture Notes in Computer Science*, 3562, 317–323.

Yang, X.S. (2008). *Nature-Inspired Metaheuristic Algorithms*. Luniver Press, Bristol.

Yang, X.S. (2010a). *Nature-Inspired Cooperative Strategies for Optimization*, Springer, Berlin, Heidelberg.

Yang, X.S. (2010b). *Engineering Optimization: An Introduction with Metaheuristic Applications*. John Wiley & Sons, New York.

Yang, X.S. (2012). Flower pollination algorithm for global optimization. *Unconventional Computation and Natural Computation, Lecture Notes in Computer Science*, 7445, 240–249.

Yang, X.S. (2013). *Artificial Intelligence, Evolutionary Computing and Metaheuristics: In the Footsteps of Alan Turing*. Springer, Berlin, Heidelberg.

Yang, X.S. and Deb, S. (2009). Cuckoo search via Lévy flights. *World Congress on Nature & Biologically Inspired Computing*, IEEE, 210–214.

Yang, X.S. and Gandomi, A.H. (2012). Bat algorithm: A novel approach for global engineering optimization. *Engineering Computation*, 29(5), 464–483.

Yang, X.S., Chien, S.F., Ting, T.O. (2014a). Computational intelligence and metaheuristic algorithms with applications. The Scientific World Journal, Article ID 425853.

Yang, X.S., Karamanoglu, M., He, X. (2014b). Flower pollination algorithm: A novel approach for multiobjective optimization. *Engineering Optimization*, 46(9), 1222–1237.

Yepes, V., Gonzalez-Vidosa, F., Alcala, J., Villalba, P. (2012). CO_2-optimization design of reinforced concrete retaining walls based on a VNS-threshold acceptance strategy. *Journal of Computing in Civil Engineering*, 26(3), 378–386.

Zhang, H.Y. and Zhang, L.J. (2017). Tuned mass damper system of high-rise intake towers optimized by improved harmony search algorithm. *Engineering Structures*, 138, 270–282.

Zhang, J.Y., Ohsaki, M., Kanno, Y. (2006). A direct approach to design of geometry and forces of tensegrity systems. *International Journal of Solids and Structures*, 43, 2260–2278.

Zhou, A., Qu, B.Y., Li, H., Zhao, S.Z., Suganthan, P.N., Zhang, Q. (2011). Multiobjective evolutionary algorithms: A survey of the state of the art. *Swarm and Evolutionary Computation*, 1(1), 32–49.

Index

Other titles from

in

Computer Engineering

2021

DELHAYE Jean-Loic
Inside the World of Computing: Technologies, Uses, Challenges

DUVAUT Patrick, DALLOZ Xavier, MENGA David, KOEHL François,
CHRIQUI Vidal, BRILL Joerg
*Internet of Augmented Me, I.AM: Empowering Innovation for a New
Sustainable Future*

2020

DARCHE Philippe
*Microprocessor 1: Prolegomena – Calculation and Storage Functions –
Models of Computation and Computer Architecture*
Microprocessor 2: Core Concepts – Communication in a Digital System
Microprocessor 3: Core Concepts – Hardware Aspects
Microprocessor 4: Core Concepts – Software Aspects
*Microprocessor 5: Software and Hardware Aspects of Development,
Debugging and Testing – The Microcomputer*

LAFFLY Dominique
TORUS 1 – Toward an Open Resource Using Services: Cloud Computing for Environmental Data
TORUS 2 – Toward an Open Resource Using Services: Cloud Computing for Environmental Data
TORUS 3 – Toward an Open Resource Using Services: Cloud Computing for Environmental Data

LAURENT Anne, LAURENT Dominique, MADERA Cédrine
Data Lakes
(Databases and Big Data Set – Volume 2)

OULHADJ Hamouche, DAACHI Boubaker, MENASRI Riad
Metaheuristics for Robotics
(Optimization Heuristics Set – Volume 2)

SADIQUI Ali
Computer Network Security

VENTRE Daniel
Artificial Intelligence, Cybersecurity and Cyber Defense

2019

BESBES Walid, DHOUIB Diala, WASSAN Niaz, MARREKCHI Emna
Solving Transport Problems: Towards Green Logistics

CLERC Maurice
Iterative Optimizers: Difficulty Measures and Benchmarks

GHLALA Riadh
Analytic SQL in SQL Server 2014/2016

TOUNSI Wiem
Cyber-Vigilance and Digital Trust: Cyber Security in the Era of Cloud Computing and IoT

2018

ANDRO Mathieu
Digital Libraries and Crowdsourcing
(Digital Tools and Uses Set – Volume 5)

ARNALDI Bruno, GUITTON Pascal, MOREAU Guillaume
Virtual Reality and Augmented Reality: Myths and Realities

BERTHIER Thierry, TEBOUL Bruno
From Digital Traces to Algorithmic Projections

CARDON Alain
Beyond Artificial Intelligence: From Human Consciousness to Artificial Consciousness

HOMAYOUNI S. Mahdi, FONTES Dalila B.M.M.
Metaheuristics for Maritime Operations
(Optimization Heuristics Set – Volume 1)

JEANSOULIN Robert
JavaScript and Open Data

PIVERT Olivier
NoSQL Data Models: Trends and Challenges
(Databases and Big Data Set – Volume 1)

SEDKAOUI Soraya
Data Analytics and Big Data

SALEH Imad, AMMI Mehdi, SZONIECKY Samuel
Challenges of the Internet of Things: Technology, Use, Ethics
(Digital Tools and Uses Set – Volume 7)

SZONIECKY Samuel
Ecosystems Knowledge: Modeling and Analysis Method for Information and Communication
(Digital Tools and Uses Set – Volume 6)

2017

BENMAMMAR Badr
Concurrent, Real-Time and Distributed Programming in Java

HÉLIODORE Frédéric, NAKIB Amir, ISMAIL Boussaad, OUCHRAA Salma, SCHMITT Laurent
Metaheuristics for Intelligent Electrical Networks
(Metaheuristics Set – Volume 10)

MA Haiping, SIMON Dan
Evolutionary Computation with Biogeography-based Optimization
(Metaheuristics Set – Volume 8)

PÉTROWSKI Alain, BEN-HAMIDA Sana
Evolutionary Algorithms
(Metaheuristics Set – Volume 9)

PAI G A Vijayalakshmi
Metaheuristics for Portfolio Optimization
(Metaheuristics Set – Volume 11)

2016

BLUM Christian, FESTA Paola
Metaheuristics for String Problems in Bio-informatics
(Metaheuristics Set – Volume 6)

DEROUSSI Laurent
Metaheuristics for Logistics
(Metaheuristics Set – Volume 4)

DHAENENS Clarisse and JOURDAN Laetitia
Metaheuristics for Big Data
(Metaheuristics Set – Volume 5)

LABADIE Nacima, PRINS Christian, PRODHON Caroline
Metaheuristics for Vehicle Routing Problems
(Metaheuristics Set – Volume 3)

LEROY Laure
Eyestrain Reduction in Stereoscopy

LUTTON Evelyne, PERROT Nathalie, TONDA Albert
Evolutionary Algorithms for Food Science and Technology
(Metaheuristics Set – Volume 7)

MAGOULÈS Frédéric, ZHAO Hai-Xiang
Data Mining and Machine Learning in Building Energy Analysis

RIGO Michel
Advanced Graph Theory and Combinatorics

2015

BARBIER Franck, RECOUSSINE Jean-Luc
COBOL Software Modernization: From Principles to Implementation with
the BLU AGE® Method

CHEN Ken
Performance Evaluation by Simulation and Analysis with Applications to
Computer Networks

CLERC Maurice
Guided Randomness in Optimization
(Metaheuristics Set – Volume 1)

DURAND Nicolas, GIANAZZA David, GOTTELAND Jean-Baptiste,
ALLIOT Jean-Marc
Metaheuristics for Air Traffic Management
(Metaheuristics Set – Volume 2)

MAGOULÈS Frédéric, ROUX François-Xavier, HOUZEAUX Guillaume
Parallel Scientific Computing

MUNEESAWANG Paisarn, YAMMEN Suchart
Visual Inspection Technology in the Hard Disk Drive Industry

2014

BOULANGER Jean-Louis
Formal Methods Applied to Industrial Complex Systems

BOULANGER Jean-Louis
Formal Methods Applied to Complex Systems: Implementation of the B Method

GARDI Frédéric, BENOIST Thierry, DARLAY Julien, ESTELLON Bertrand, MEGEL Romain
Mathematical Programming Solver based on Local Search

KRICHEN Saoussen, CHAOUACHI Jouhaina
Graph-related Optimization and Decision Support Systems

LARRIEU Nicolas, VARET Antoine
Rapid Prototyping of Software for Avionics Systems: Model-oriented Approaches for Complex Systems Certification

OUSSALAH Mourad Chabane
Software Architecture 1
Software Architecture 2

PASCHOS Vangelis Th
Combinatorial Optimization – 3-volume series, 2nd Edition
Concepts of Combinatorial Optimization – Volume 1, 2nd Edition
Problems and New Approaches – Volume 2, 2nd Edition
Applications of Combinatorial Optimization – Volume 3, 2nd Edition

QUESNEL Flavien
Scheduling of Large-scale Virtualized Infrastructures: Toward Cooperative Management

RIGO Michel
Formal Languages, Automata and Numeration Systems 1: Introduction to Combinatorics on Words
Formal Languages, Automata and Numeration Systems 2: Applications to Recognizability and Decidability

SAINT-DIZIER Patrick
Musical Rhetoric: Foundations and Annotation Schemes

TOUATI Sid, DE DINECHIN Benoit
Advanced Backend Optimization

2013

ANDRÉ Etienne, SOULAT Romain
The Inverse Method: Parametric Verification of Real-time Embedded Systems

BOULANGER Jean-Louis
Safety Management for Software-based Equipment

DELAHAYE Daniel, PUECHMOREL Stéphane
Modeling and Optimization of Air Traffic

FRANCOPOULO Gil
LMF — Lexical Markup Framework

GHÉDIRA Khaled
Constraint Satisfaction Problems

ROCHANGE Christine, UHRIG Sascha, SAINRAT Pascal
Time-Predictable Architectures

WAHBI Mohamed
Algorithms and Ordering Heuristics for Distributed Constraint Satisfaction Problems

ZELM Martin *et al.*
Enterprise Interoperability

2012

ARBOLEDA Hugo, ROYER Jean-Claude
Model-Driven and Software Product Line Engineering

BLANCHET Gérard, DUPOUY Bertrand
Computer Architecture

BOULANGER Jean-Louis
Industrial Use of Formal Methods: Formal Verification

BOULANGER Jean-Louis
Formal Method: Industrial Use from Model to the Code

CALVARY Gaëlle, DELOT Thierry, SÈDES Florence, TIGLI Jean-Yves
Computer Science and Ambient Intelligence

MAHOUT Vincent
Assembly Language Programming: ARM Cortex-M3 2.0: Organization, Innovation and Territory

MARLET Renaud
Program Specialization

SOTO Maria, SEVAUX Marc, ROSSI André, LAURENT Johann
Memory Allocation Problems in Embedded Systems: Optimization Methods

2011

BICHOT Charles-Edmond, SIARRY Patrick
Graph Partitioning

BOULANGER Jean-Louis
Static Analysis of Software: The Abstract Interpretation

CAFERRA Ricardo
Logic for Computer Science and Artificial Intelligence

HOMES Bernard
Fundamentals of Software Testing

KORDON Fabrice, HADDAD Serge, PAUTET Laurent, PETRUCCI Laure
Distributed Systems: Design and Algorithms

KORDON Fabrice, HADDAD Serge, PAUTET Laurent, PETRUCCI Laure
Models and Analysis in Distributed Systems

LORCA Xavier
Tree-based Graph Partitioning Constraint

LECOUTRE Christophe
Constraint Networks / Targeting Simplicity for Techniques and Algorithms

2008

BANÂTRE Michel, MARRÓN Pedro José, OLLERO Hannibal, WOLITZ Adam
Cooperating Embedded Systems and Wireless Sensor Networks

MERZ Stephan, NAVET Nicolas
Modeling and Verification of Real-time Systems

PASCHOS Vangelis Th
Combinatorial Optimization and Theoretical Computer Science: Interfaces and Perspectives

WALDNER Jean-Baptiste
Nanocomputers and Swarm Intelligence

2007

BENHAMOU Frédéric, JUSSIEN Narendra, O'SULLIVAN Barry
Trends in Constraint Programming

JUSSIEN Narendra
A TO Z OF SUDOKU

2006

BABAU Jean-Philippe *et al.*
From MDD Concepts to Experiments and Illustrations – DRES 2006

HABRIAS Henri, FRAPPIER Marc
Software Specification Methods

MURAT Cecile, PASCHOS Vangelis Th
Probabilistic Combinatorial Optimization on Graphs

PANETTO Hervé, BOUDJLIDA Nacer
Interoperability for Enterprise Software and Applications 2006 / IFAC-IFIP I-ESA'2006

2005

GÉRARD Sébastien *et al.*
Model Driven Engineering for Distributed Real Time Embedded Systems

PANETTO Hervé
Interoperability of Enterprise Software and Applications 2005

Printed and bound by CPI Group (UK) Ltd, Croydon, CR0 4YY